DEEP CUT

SINCE1970
Histories of Contemporary America

Deep Cut

Science, Power, and the Unbuilt Interoceanic Canal

Christine Keiner

THE UNIVERSITY OF GEORGIA PRESS

ATHENS

Most University of Georgia Press titles are available from popular e-book vendors.

Library of Congress Control Number: 2020932724
ISBN: 9780820358635 (ebook: open access edition)
ISBN: 9780820338941 (hardback: alk. paper)
ISBN: 9780820338958 (paperback: alk. paper)
ISBN: 9780820358307 (ebook: standard edition)

An earlier version of material from chapters 1 and 3 appeared, in very different form, within Ashley Carse, Christine Keiner, Pamela M. Henson, Marixa Lasso, Paul S. Sutter, Megan Raby, and Blake Scott, "Panama Canal Forum: From the Conquest of Nature to the Construction of New Ecologies," *Environmental History* 21 (2016): 206–87. An earlier version of material from chapters 5–6, in very different form, appeared in Christine Keiner, "A Two-Ocean Bouillabaisse: Science, Politics, and the Central American Sea-Level Canal Controversy," *Journal of the History of Biology* 50 (2017): 835–87, to which Springer Nature retains copyright.

S | H
M | P

The Sustainable History Monograph Pilot

Opening up the Past, Publishing for the Future

This book is published as part of the Sustainable History Monograph Pilot. With the generous support of the Andrew W. Mellon Foundation, the Pilot uses cutting-edge publishing technology to produce open access digital editions of high-quality, peer-reviewed monographs from leading university presses. Free digital editions can be downloaded from: Books at JSTOR, EBSCO, Hathi Trust, Internet Archive, OAPEN, Project MUSE, and many other open repositories.

When you cite the book, please include the following URL for its Digital Object Identifier (DOI): https://doi.org/10.46935/9780820358635

We are eager to learn more about how you discovered this title and how you are using it. We hope you will spend a few minutes answering a couple of questions at this url: **https://www.longleafservices.org/shmp-survey/**

More information about the Sustainable History Monograph Pilot can be found at https://www.longleafservices.org.

To the Memory of James A. Keiner

1940–2016

Sailor, Civil Servant, Father, Friend

CONTENTS

ILLUSTRATIONS

FIGURES

MAPS

ACKNOWLEDGMENTS

I am grateful beyond words to my generous mentors, friends, and colleagues for helping me in many ways: Sharon Kingsland, Pamela Henson, Jeffrey Stine, James Carlton, Robert Kargon, Helen Rozwadowski, Kurk Dorsey, Ronald Doel, Stephen Bocking, Ashley Carse, Marixa Lasso, Megan Raby, Blake Scott, Paul Sutter, Penelope Hardy, Lincoln Paine, John Cloud, Matthew Booker, Gerard Fitzgerald, Matt Chew, Roger Turner, Jeremy Vetter, Mark Hersey, Matt McKenzie, Daniel Macfarlane, Scott Kaufman, Mark Lawrence, Shaine Scarminach, Katey Anderson, Jordan Coulombe, Derek Nelson, Jake Hamblin, Tony Adler, Samantha Muka, Karen Rader, JoAnn Palmeri, Katie Terezakis, Tamar Carroll, Rebecca Edwards, Michael Laver, Rich Newman, Rebecca Scales, Corinna Schlombs, Rebecca DeRoo, Deborah Blizzard, Tom Cornell, Ann Howard, Kristoffer Whitney, Sandra Rothenberg, LaVerne McQuiller Williams, and James Winebrake.

I deeply appreciate the work of the University of Georgia Press's team of editors, managers, and production associates, both past and present, especially Mick Gusinde-Duffy, Derek Krissoff, Andrew Berzanskis, Lynn Itagaki, Daniel Rivers, Jon Davies, Beth Snead, David Des Jardines, Sara Ash Georgi, Erin Kirk, Ihsan Taylor of Longleaf Services, and two extremely helpful anonymous reviewers.

To the organizers and audience members of seminars at which I presented earlier versions of this work, thank you for your hospitality and helpful feedback: Tom Lassman of the Smithsonian History Seminar on Contemporary Science and Technology; Eric Roorda, Glenn Gordinier, and Carol Mowrey of the Munson Institute of American Maritime Studies at Mystic Seaport; Betsy Mendelsohn and David Kirsch of the University of Maryland Colloquium in the History of Technology, Science, and Environment; Zachary Cuyler and Troy Vettese of the New York University Energy and the Left Workshop; the Johns Hopkins University Department of the History of Science and Technology Colloquium; the Smithsonian Tropical Research Institute Barro Colorado Island Bambi Seminar; the Tri-University History Conference on Cold War Encounters; and the University of Connecticut–Avery Point Coastal Perspectives

Lecture Series. Thanks also to the organizers, panelists, and audience members of conference sessions at the American Association of Geographers, American Historical Association, American Society for Environmental History, Columbia History of Science Group, History of Science Society, North American Society for Oceanic History, Rochester U.S. Historians, Society for the History of American Foreign Relations, and Society for the History of Technology.

Many thanks to Ira Rubinoff, Alan Covich, Wayne Clough, William Newman, and other historical participants for sharing their insights with me. Any mistakes are of course my own.

For their generous hospitality in Panama, muchas gracias to Noris Herrera, Susan Brewer-Osorio, Ariel Espino, Dan Norman, Stanley Heckadon-Moreno, Egbert Leigh, Rachel Collin, Héctor Guzmán, Harilaos Lessios, and Mark Torchin.

Many archivists, librarians, and interlibrary loan officers provided crucial sources, for which I am most appreciative. Thank you also to the program officers and support staff of the Smithsonian Institution Archives Postdoctoral Fellowship Program, National Endowment for the Humanities Summer Stipend Program, Lyndon B. Johnson Foundation Moody Research Grant Program, and Eisenhower Foundation Abilene Travel Grants Program for providing critical funds and for believing in my project. I am also very grateful to the Rochester Institute of Technology College of Liberal Arts Miller Fellowship, Faculty Research Fellowship, and Publication Cost Grant Programs, and to the RIT Departments of STS and History.

Deep thanks to my fantastic high school teachers and college professors for setting high standards of mentorship to which I have always aspired, especially Zeleana Morris, Rod Wallace, Kenneth Zachmann, Esther Iglich, Christianna Nichols Leahy, and Carole McCann.

I am very blessed by my supportive family: Sonia, Matt, AJ, Greg, Dana, Ethan, Andrew, Samuel, Gabriel, Gary, Helen; my aunts, uncles, cousins, and late grandparents; and my wonderful mom, Vera. Thank you ADK for the loan of many history books! And the greatest thanks of all to my dear and loving husband, Darren Lacey. This book is dedicated in memory of Jim Keiner, taken too soon from us all by pancreatic cancer.

Introduction

The Central American Sea-Level Canal and the Environmental History of Unbuilt Megaprojects

THE ATLANTIC-PACIFIC Central American sea-level canal was a spectacular failure. The famed French engineer of the Suez Canal, Ferdinand de Lesseps, destroyed his career and the lives of twenty-five thousand workers by insisting he could excavate across the mountainous Panamanian isthmus in the 1880s. Learning from his mistakes, the Americans succeeded in the early 1900s by taming the tropical insect-borne diseases and building an ingenious "bridge of water" with a dam and locks that lift ships eighty-five feet above sea level. However, the world-changing Panama Canal appeared more and more obsolescent and vulnerable as ship size and military airpower increased, leading its postwar operators to dust off old plans for a means of connecting the seas without any impediments. Nuclear weapons designers thought they had discovered the key to unlocking the canal in the form of peaceful nuclear explosives (PNEs), and to that end the U.S. government spent hundreds of millions of dollars and thirteen years considering the question of atomic excavation. And yet, when a presidential commission announced in 1970 that it had ruled out PNEs in favor of ordinary construction methods, critics from both the right and the left in essence declared, "told you so."

That in a nutshell is the conventional story of the Atlantic-Pacific sea-level canal, a megaproject that failed to make the transition from idea to reality and thereby enter the pantheon of monumental civil engineering works. Devoting historiographical attention to such a nonevent might seem counterproductive.[1] But over the last two decades scholars have produced insightful technoscientific and diplomatic histories of the nuclear canal proposal (de Lesseps's project has received more attention, though mainly as a counterpoint to the American success). Scholarly studies of the nuclear canal have enriched our understanding of Cold War–era mentalities and geopolitical relations by addressing it in the context of Project Plowshare, the 1957–77 U.S. initiative to apply nuclear energy to earthmoving and other nonmilitary pursuits.[2] The "Panatomic" proposal sheds light not only on the hubris and tenacity of Plowshare, but also the arrogance and persistence of U.S. imperialism in Panama, which lasted from 1903 (when the United States helped engineer a revolt against Colombia) to 1999 (when Panama assumed full sovereignty over the waterway and surrounding zone).[3]

This book examines the nuclear canal in a different context, as one of several science-based iterations of an anticipated infrastructural future that began with the founder of environmental science, Alexander von Humboldt, and stretched to the turn of the millennium, almost three decades after the demise of Plowshare. Although the sea-level ship canal did not come to pass, as a *proposal* it served important political and scientific purposes during different eras. During the 1950s and 1960s, it enabled three U.S. presidents to address the increasingly problematic Panama Canal Zone, the colonialistic enclave surrounding the original waterway, and during the 1970s and 1980s, it offered new visions for dealing with the oil crisis. Throughout the most serious period of governmental attention, from 1965–70, the plan provided opportunities for producing new knowledge to resolve the burning question of whether detonating buried thermonuclear bombs to excavate the "very deep cuts required by an interoceanic canal" was technically feasible.[4] The U.S. government's then lack of legal requirements for assessing environmental impacts, and its narrow definition of the kinds of information, expertise, and authoritative capacity deemed most relevant to this task, in turn generated high-profile debates within the scientific community over the project's nonradioecological, nonanthropocentric risks. After officials ruled out nuclear construction methods—an outcome that was never inevitable—ecological concerns about how the sea-level canal and other maritime transportation complexes might affect the biological integrity of the oceans continued to circulate in the policy realm in ways that still resonate today.

In other words, as science and technology studies scholars would say, the sea-level canal proposal performed different kinds of work for different historical actors, and vice versa.[5] Dismissing it as a failed scheme unworthy of historiographical attention prevents us from considering the political, cultural, and epistemological processes that went into constructing the seaway as an innovative diplomatic solution, an exciting research opportunity, a superior hydrocarbon highway, or a serious ecological threat. These processes highlight themes of broader relevance to environmental history and the history of science and technology. The controversies over the nuclear and nonnuclear phases of the sea-level canal signify the disintegration of a powerful technocratic worldview that permeated U.S. environmental management from the construction of the original Panama Canal until the rise of the modern environmental movement.

Deep Cut uses the Central American seaway proposal to address specific elements and consequences of this revolutionary shift, especially the changing roles of environmental expertise and state-sponsored preliminary environmental impact assessment, and the historical contingencies of infrastructural decision-making. More broadly, this book contributes to an emerging literature addressing the environmental, scientific, and political histories and legacies of unbuilt megaprojects.

AS SUGGESTED BY a small but growing number of researchers from diverse fields, the history of planned but unrealized projects—from single buildings to regional development endeavors—deserves much greater attention. One of the first studies of "unbuiltism" appeared over forty years ago. An art historian coined this odd word in the introductory essay of *Unbuilt America*, a book featuring dozens of salvaged drawings and descriptions uncovered by two artists from architectural firms, libraries, and private collections in the 1970s.[6] According to the authors, only one other such work dealing "exclusively with the unbuilt as a phenomenon" had ever appeared, and that was in 1925.[7] *Unbuilt America* focuses on buildings and monuments, but the reasons proffered for their lack of execution apply as well to larger-scale architectural, engineering, and landscape projects. Visions of the built environment fail to take form for many reasons, including lack of funding, bureaucratic inertia, technical infeasibility, and community resistance. The book did not unleash a groundswell of academic or popular interest in unbuiltism, but its time has come. A recent set of coffee-table books devoted to never-built structures envisioned for Los Angeles and New York have become best sellers and spawned Kickstarter-funded museum exhibits meant to inspire viewers to consider what might have been and what might be different for the human-dominated landscapes of the future.[8]

The curators of the Never Built series hail from the world of architecture, but more deeply contextualized case studies of unrealized large-scale projects have emerged from across the environmental humanities.[9] Cultural historian Kathryn Oberdeck coined the phrase "unbuilt environment" in a 2005 essay about the value of archives as conservatories of unrealized urban visions.[10] Moreover, in the words of geographer Michael Heffernan, "Unsuccessful initiatives, especially controversial and long-running ones, tend to leave an archival legacy that is more complex and extensive than realized projects. Failures allow the historian to chart the limits of our faith in science and technology." Analysis of the private and public evidence of grandiose ventures that never came about can also shed light on imperialist and modernist attitudes that continued to influence development thinking long afterward. These are important points of Heffernan's pioneering studies of late nineteenth-century French colonial plans to transform the Sahara Desert into a vast inland sea and railway network that would fuel the development of northern Africa.[11]

Another never-built Eurocentric macroengineering project with rich insights for the history of science, technology, and human-environment relations was Atlantropa. From 1927 until his death in 1952, German architect Herman Sörgel sought to overcome European fears of energy shortages and cultural decline by damming the Mediterranean to generate hydroelectricity and lower the sea enough to create a new continent connecting Europe and Africa. In his vision, climate engineering would stave off desertification, thereby ensuring healthful conditions for Atlantropa's imperialist European communities and industries. The megaproject's scale, complexity, and cost undermined its feasibility, but its failure was not preordained. The idea circulated in German academic and popular media for over two decades, revealing "lesser-known environmental issues and fears in the first half of the twentieth century that—in altered form—are still with us today." Atlantropa elucidates "the still-understudied history of the unrealized utopian projects of high modernism."[12]

High modernism refers to a philosophy underpinning massive projects of the twentieth century that came to fruition but otherwise failed to achieve their lofty goals of improving the human condition, in the words of political scientist and anthropologist James Scott. Scott's influential analysis uses cases such as the Soviet collectivization of agriculture and the Brazilian construction of a new capital city in the Amazon rainforest to demonstrate how high modernists mobilized science and technology to advance progress, yet wound up causing great ecological and social harm by privileging centralized technocratic expertise and rigid centralized directives above local knowledge and needs.[13]

The most notorious high-modernist projects have taken place in authoritarian societies capable of crushing dissent, but democratic governments have also promoted problematic large-scale agricultural, industrial, and urban development projects despite fierce local resistance. A revelatory example is the St. Lawrence Seaway, the U.S.-Canadian transportation and hydroelectricity complex that submerged several villages and displaced 6,500 citizens. Historian Daniel Macfarlane uses the concept of "negotiated high modernism" to show how U.S. and Canadian officials strategically used the public planning process to overcome opposition from stakeholders who did not stand to benefit. Decision makers had to "adapt, negotiate, and manufacture consent in order to achieve a veneer of democratic legitimacy" to actualize their vision.[14] This concept is also useful for examining technocratic ventures that did not pass the regulatory policy-making processes of liberal democracies.

One of the ultimate high-modernist enterprises that did not advance beyond the experimental stage, and to which the concept of negotiated high modernism applies, was the aforementioned Project Plowshare program of the U.S. Atomic Energy Commission (the Soviet Union operated its own such program for even longer). Starting in 1957, Plowshare officials convinced Congress that investing in PNE feasibility studies would reap huge dividends in the form of cost-effective transportation infrastructure (harbors, roads, and canals), energy sources (especially the extraction of oil and gas from shale and tar sands), medical breakthroughs (via the production of isotopes), and weather modification.[15] As for the question of releasing radioactivity into the environment, Plowshare scientists and engineers, most of whom worked at the Lawrence Radiation Laboratory at the University of California at Livermore, believed they could develop what they called clean explosives. Because a PNE consisted of a thermonuclear device with a fission trigger, the smaller the amount of harmful radionuclides released via the fission process and the deeper the device was buried, the less the radiation hazard. However, placing a PNE too far underground would reduce its effectiveness for earthmoving purposes. Figuring out the optimal combinations of such variables as burial depth, percentage of vented radiation, and the number of explosives needed for a given goal required extensive testing. The Atomic Energy Commission (AEC) conducted twenty-seven experimental explosions for Plowshare, primarily at the Nevada Test Site, from 1961 to 1973. Each one required presidential permission and intensive calculating of political risks, especially after the Limited Nuclear Test Ban Treaty of 1963 banned countries conducting atomic experiments from releasing radioactive debris across national borders.

Today the idea of detonating buried hydrogen bombs to excavate a waterway or frack natural gas seems absurd if not "mildly deranged."[16] Plowshare's most famous supporter was the physicist Edward Teller, whose uncompromising advocacy for the hydrogen bomb might have inspired the 1964 movie *Dr. Strangelove*.[17] Likewise, Teller's plans to reshape the earth with PNEs now make for compelling clickbait; recent media accounts of Plowshare feature adjectives such as *bizarre*, *crazy*, *insane*, *madcap*, *outlandish*, and *wacky*.[18]

But for a confident, technocratic society seeking to accelerate modernity, PNEs were anything but bizarre—rather, they held the promise of harnessing knowledge to enhance the quality of life at bargain prices. Federal officials were not required to take what we would now call environmental costs into consideration, which helps explain why for them nuclear excavation appeared much less expensive than conventional methods. As Teller and his allies argued, atomic excavation would not only save money but also reduce the number of worker injuries and deaths caused by massive construction projects. For such reasons Plowshare advocates considered themselves "the personification of progress and modernity."[19] For scholars of high modernism, of the Cold War, and of the history of science, technology, and the environment, Project Plowshare offers a window into a worldview that revered technoscience to the point of assuming that serious environmental health risks could be contained and that dissent on such grounds was irrational.

Like many midcentury Americans, Plowshare's powerful sponsors in Congress and the executive branch held great faith in the power of science and technology to solve social and political problems. At the same time, U.S. decision makers operated in the context of checks and balances. However imperfectly, unevenly, and inconsistently, democratic governing systems provide opportunities for challenge and change. Examining how proponents of Plowshare dealt with outsiders asking tough environmental and public health questions shines light on the evolving political role of scientific expertise and dissent vis-à-vis environmental impact assessment in the mid-to-late twentieth century.[20]

Plowshare's first proposed venture, Project Chariot, aimed to excavate a harbor in northwestern Alaska. Deep historical detective work has exposed the secretive, hubristic ways in which Livermore's scientist-administrators tried to exploit the Alaskan tundra for their own purposes, especially to sell the Central American sea-level canal and thereby keep the laboratory running while the U.S. observed a voluntary moratorium on nuclear weapons testing from 1958–61.[21] Another theme of this literature is the rise of antinuclear environmental advocacy. In response to unexpected pushback from Indigenous people, local biologists, and citizen conservationists regarding radiation hazards, Teller and his Livermore associates sought to co-opt the opposition by sponsoring extensive bioenvironmental studies.[22] However, rather than meeting AEC expectations, the AEC-funded University of Alaska biologists joined with grassroots allies to publicize both their pioneering research and their political arguments against the proposal.[23]

The Chariot studies and publicity had far-reaching effects. Barry Commoner, the biologist-activist and originator of the famous quote, "The first law of ecology is that everything is related to everything else," traced his ecological awakening to the insight that Arctic lichen would absorb radioactive fallout from the atmosphere and contaminate the caribou grazing on them, in turn harming the human carnivores atop the tundra food chain.[24] Moreover, the final official bioenvironmental report, which came out in 1966, four years after the AEC responded to the public uproar by canceling Chariot, has been hailed as a model for the first modern environmental impact statement.[25]

But getting to that point took much work and maneuvering by Chariot's conflicting interests: "It was precisely through resistance to Plowshare program plans—and through Livermore's strategies for *overcoming* resistances—that the environmental program took shape as it did."[26] Teller and his allies spent half a decade negotiating their high-modernist vision of plowsharing an instant harbor, and in the process wound up fostering new conceptions of progress, environmental and human health risk assessment, and planning in a high-tech society.[27]

That is not to say that such conceptions quickly transformed statist environmental management, nor that the AEC implemented such insights as it pursued PNE projects in other places over the next several years: "It was notoriously slow in learning lessons," especially about overcoming public resistance to conducting experimental blasts outside the confines of the Nevada Test Site.[28] Like others at the forefront of technological innovation, AEC and Plowshare administrators exhibited "uniqueness bias," the tendency of planners to "see their projects as firsts, which impedes learning from other projects."[29] Despite the opposition of Alaskans to Project Chariot, a decade later the AEC persisted with an unpopular test in Colorado to investigate the feasibility of nuclear fracking. Although citizen environmentalists failed to prevent the detonation from taking place in October 1969, their groundbreaking lawsuit subjected the AEC to judicial review.[30] The subsequent passage by Congress of the National Environmental Policy Act of 1969 (NEPA) provided Plowshare opponents with an even more potent legal instrument. By requiring federal infrastructure designers to conduct preliminary environmental impact studies, articulate less damaging alternatives, and solicit comments from the public, NEPA changed public works planning in the last three decades of the twentieth century in the United States, as well as in other democratic nations.[31]

NEPA and other environmentalist initiatives blunted the high-modernist and utilitarian rationales that had characterized the broader complex within which the AEC and other federal agencies operated to help citizens control nature. Historians of the environmental management state have begun to explicate how U.S. practices and policies designed to overcome environmental challenges functioned and evolved as part of a systematic enterprise of state building.[32]

Indeed, the Panama Canal itself constitutes a key case study of how U.S. officials solved complex environmental and public health problems by mobilizing science and technology on a scale comparable to waging war.[33]

Within six decades of the canal's construction, however, state-sponsored environmental management underwent major transformations: from a confident, engineering-dominated enterprise in the heroic service of prosperity and modernity to a more ambiguous endeavor requiring recognition of uncertainty, sensitivity to multiple and nonelite stakeholders, consideration of alternatives, and suspicion of high-modernist technological solutions. By the 1970s, as knowledge of complex ecological consequences expanded and gained political legitimacy in democratic societies via laws like NEPA, civil engineers and hydro-engineers could no longer presume to exert the same levels of control as had their predecessors.[34] Accordingly, addressing how statist environmental management policies and practices shifted over the middle decades of the twentieth century from conquering nature to implementing less damaging forms of development provides another rationale for paying attention to unrealized macroengineering projects and the debates they generated.

Toward this end, environmental historians have delineated several important controversial proposals beyond the realm of nuclear power that withered in the face of intense, sustained resistance. Recent studies of large-scale energy production and transportation plans that did not materialize during the 1960s and 1970s emphasize the skillful mobilization of scientific knowledge and new legal tactics by grassroots groups. U.S. environmentalists, often though not always allied with scientists, stopped the Storm King Mountain hydroelectric storage facility, designed to supply power for New York City; the Cross Florida Barge Canal, intended to provide a maritime shortcut across northern Florida; the completion of the Pan-American Highway through the Darién Gap of Panama and Colombia, a U.S. foreign aid project; and the U.S. supersonic transport aircraft program and associated Everglades Jetport, which was designed to be the world's largest airport.[35] Local environmentalist opposition and high costs also contributed to the early-1970s demise of Scripps Island, a planned underwater research facility on which the prestigious Scripps Institution of Oceanography spent almost a decade. Its downfall tells us much not only about how rapidly citizens in California and elsewhere redefined the appropriate use of marine and coastal resources, but also about changing scientific attitudes toward the practice and goals of oceanography.[36]

Other recent publications provide a rich foundation for rethinking unbuilt projects. Geographer Jonathan Peyton's work on the history of unrealized development in a remote corner of British Columbia urges us to consider the "capacities that are created by failed schemes" and the "conditions of possibility" generated by such plans.[37] How did the discourse related to development proposals—including the data produced to justify them—affect both the physical environment and human perceptions of it? Drawing upon decades' worth of unsuccessful plans to industrialize the Stikine region of northwestern Canada, Peyton shows how such failures reinforced rather than weakened perceptions of the Stikine as a resource-rich landscape destined for development. Previous unsuccessful attempts to mobilize the area's resources on a grand scale continue to exert influence on contemporary controversies regarding "the politics and possibilities of development" because they created things—paperwork, knowledge claims, memories, and physical landscape imprints such as geographical surveys—with which current stakeholders must reckon if they are to triumph in the latest round of debate. "The question is not why did something fail but, rather, what did failure produce, what kinds of effects did it have?" Examining what Peyton calls the multilayered side effects of unsuccessful development proposals accentuates the historical contingency of infrastructure project planning—past, present, and future.[38]

In a parallel vein, this book explores several multilayered effects that emerged over time as different U.S. actors invoked the isthmian sea-level canal proposal, and its exciting conditions of possibility, to solve different problems. For each of these contexts, it is useful to ask: What kinds of political and epistemic consequences did the proposal have, and what do they tell us about the changing political role of scientific expertise in the context of state-sponsored environmental management, especially regarding the concept of preliminary impact assessment? And how did unexpected obstacles posed by changing economic, political, and cultural climates, as well as the physical environment itself, alter the expectations and strategies of the proposal's patrons? Addressing these questions invites us to interweave insights from multiple disciplines, especially diplomatic history and the environmental history of science, an approach that recognizes nature as an active shaper of scientific knowledge and institutions, not only as a passive subject of scientific inquiry.[39]

Seen through these lenses, the sea-level canal saga illuminates how geopolitical imperatives facilitated specific kinds of preproject field research in politically sensitive places, and in turn, how meteorological, geological, and other environmental constraints and discoveries subverted political and scientific expectations of how those landscapes could be developed. Collectively, the disparate phases of the failed seaway endeavor demonstrate how convergences of unforeseen political, economic, military, technological, scientific, and environmental forces sometimes stymied high-modernist plans, while also setting the stage for later iterations of the envisioned infrastructure.[40]

The first chapter of part I sets the scene by providing an alternative "shadow history" of the Panama Canal creation story that foregrounds the role of the sea-level design in the long quest to link the Atlantic and Pacific Oceans. Shadow histories utilize primary sources and other remnants of forgotten debates to remind us of visions of the future that once seemed possible.[41] In this case, Ferdinand de Lesseps's ruinous 1889 failure to cut all the way down across the isthmus should not eclipse the Euro-American scientific enthusiasm underpinning the endeavor. Indeed, the most famous scientist of the century, Alexander von Humboldt, who spent decades advocating for an artificial waterway, explicitly endorsed an interoceanic canal without obstructive locks in the 1850s.[42] Moreover, for two years after the U.S. government resumed digging in 1904, arguments continued over whether to adopt the lock design that won out in the end.

Despite the triumphant opening of the lock canal in 1914, problems predicted by the sea-level advocates soon began to materialize. Chapters 2 and 3 address the growing technological obsolescence of the Panama Canal during the interwar years, the Atomic Age rationales and techniques for excavating a new waterway, the increasing Panamanian resistance against U.S. control of the Panama Canal and its colony-like enclave, and finally, the 1964 diplomatic crisis that led President Lyndon Johnson to announce that the United States would "plan in earnest" to replace the existing canal and renegotiate the inequitable 1903 treaty with Panama.[43] For Johnson and his two predecessors, the sea-level canal offered a means of driving new discussions over decolonizing the Canal Zone while maintaining U.S. hegemony in the Western Hemisphere. In intriguing ways, the seaway proposal functioned as an instrument of both technological imperialism and the far rarer phenomenon of technological *anti*-imperialism.[44]

Part II transitions from the political opportunities that the nuclear canal idea provided for U.S. leaders to the preproject feasibility studies and ensuing debates. Focusing on the civilian presidential appointees of the Atlantic-Pacific Interoceanic Canal Study Commission, rather than Teller and the Livermore physicists, opens new interpretive pathways on the Panatomic Canal endeavor. The commissioners, whose backgrounds included law, engineering, administration, and the military, were enthusiastic about PNEs, but not blindly so. From 1965 to 1970 they oversaw a wide-ranging investigation, and when their final report recommended proceeding with sea-level canal construction using nonnuclear methods, critics across the political spectrum vilified them for wasting millions of dollars and minimizing scientific concerns about an emerging environmental issue: marine biological invasions. However, their declassified meeting transcripts reveal the flexibility with which the commissioners strove to complete their complicated mission, especially in the face of severe budgetary, meteorological, and time constraints and pressure from the scientific community that challenged their high-modernist tendencies. The commission's work helps us appreciate the evolution of an adaptively technocratic form of preliminary environmental impact assessment in the transformative decade between Project Chariot and NEPA.

Toward this end, chapter 4 examines how the Canal Study Commission navigated a difficult course set by Congress and the president to determine the proposed waterway's nuclear feasibility and optimal location. The commission's workload included overseeing the collection of reams of economic, military, and engineering data and analyses of the seaway's feasibility. At the same time, the chairman endeavored to persuade reluctant U.S. officials to allow the needed PNE experiments to take place at the Nevada Test Site, in spite of the risk of violating the Limited Nuclear Test Ban Treaty provision against cross-border radioactive debris. The commission chairman also led the 1964–67 initiative to renegotiate the 1903 treaty with Panama, a process in which the seaway proposal played a large part. Multiple diplomatic contingencies affected the scientific and engineering field studies, and vice versa.

Chapters 5 and 6 address how the commissioners dealt with scientific and political developments that destabilized the U.S. government's high hopes for the proposed nuclear seaway. The engineering feasibility studies, which took place in the remote Darién portions of eastern Panama and northwestern Colombia, entailed numerous diplomatic and technical difficulties. The research teams struggled to complete their work in the contexts of the tropical climate, which limited operations to the four-month dry season, and the Vietnam War, which diverted needed funds, equipment, and personnel. The high-modernist dream of plowsharing the Darién collapsed in a literal and political quagmire.

A second set of challenges unfolded in a much more public way, as biologists associated with the Smithsonian Institution and several universities called for a share of the engineering feasibility funds to address the nonanthropocentric, nonterrestrial matter of marine species exchange. Ecological and evolutionary researchers raised concerns about the sea-level canal's ability to allow sea snakes and other invasive organisms to cross between the Pacific and the Atlantic on a massive scale for the first time in approximately three million years. The issue led to heated debates between biologists and engineers about the oceans' biological integrity, and among scientists about whether the megaproject represented a research opportunity or environmental threat. Biologists cannot take credit for stopping the nuclear canal, but they did contribute to new understandings of the areas of expertise considered relevant for assessing the environmental risks of maritime infrastructure.

Part III explores a subsequent side effect of the sea-level canal proposal, its nonnuclear revival by President Jimmy Carter as a solution to the 1970s-era oil crisis. An Alaskan senator convinced the president that a sea-level channel would facilitate the flow of North Slope oil to East Coast, Caribbean, and Gulf Coast refineries and thereby preclude the construction of expensive, disruptive new networks of terrestrial pipelines. As discussed in chapter 7, Carter confounded his environmentalist and diplomatic allies by insisting that the 1977 Panama Canal Treaty include a provision allowing a U.S. option for a new sea-level waterway. The last-minute addition created new kinds of work for the federal agencies required by NEPA to assess the sea-level canal's likely environmental impacts. The treaty clause also rallied environmental advocacy groups, whose influence had grown significantly during the so-called environmental decade inaugurated by President Richard Nixon's signing of the National Environmental Policy Act on January 1, 1970.

In this post-NEPA era, as addressed in chapter 8, the sea snake studies of the 1960s acquired a new political significance. Both conservative opponents of the Panama Canal Treaties and professionalized environmentalist organizations mobilized marine invasion ecology to discredit the seaway proposal. Moreover, the State Department incorporated scientific and political insights from the earlier canal ecology controversy into two remarkable documents: a groundbreaking environmental impact study of the 1977 pact granting Panama sovereignty over the waterway and surrounding enclave, and a statement of reassurance designed to win domestic environmentalist support for the treaty ratification campaign of 1978. This phase of the sea-level canal controversy signified the dissolution of the technocratic worldview that had underpinned the Panatomic proposal in favor of more democratic and precautionary modes of environmental governance.

The book concludes with a brief discussion of how Japanese interests revived the seaway idea in the 1980s for importing hydrocarbons and how more recent events link back to the feasibility studies of the late 1960s. By abrogating the 1903 accord, the 1977 canal treaties set the stage for a new diplomatic relationship in which Panama transitioned over the course of two decades toward assuming full responsibility for managing the waterway. The nation's decision to modernize it by building a wider lane of locks from 2007 to 2016 drew in some ways on the ecological insights of the sea-level canal debates of the late 1960s and 1970s. Although environmentalists criticized Panama for not conducting a thorough, democratically informed environmental impact assessment, the canal now features a water-saving mechanism that has so far precluded the need to pump in seawater, which, as biologists warned over four decades ago, would eliminate a crucial freshwater barrier to marine species exchange.

Moreover, the biologists who put pressure on President Johnson's Canal Study Commission set an example that has influenced recent scientific responses to the world's largest planned infrastructural excavation project, the 170-mile-long Nicaragua Grand Canal.[45] In 2014, the Nicaraguan government announced that construction had already begun, a year and a half after awarding the $40 billion concession to a Chinese business magnate eager to develop an alternative lock-based ship channel. Although the Nicaraguan government responded to protests by green-lighting an environmental and social impact assessment, a panel of international experts found it lacking. The country's highest court dismissed the last environmentalist challenges in 2017, yet the project remains in limbo, a likely consequence of the Chinese concessionaire's economic downfall as well as improved China-Panama relations. Like the Panama Canal expansion, the Nicaragua plan sheds light on how twenty-first-century Latin American nations weigh development priorities with post-NEPA concepts of statist environmental governance and public input procedures, especially in the context of the growing power of Asian energy and maritime shipping interests.

Ending with a yet-unrealized megaproject, especially one that embraces a resurgent form of high modernism, reiterates one of the book's undercurrents: the idea that in order to come to fruition, large-scale infrastructure ventures require multiple forces—political, economic, technological, scientific, and environmental—to align in just the right way at just the right time. Tracing such developments, along with challenging entrenched assumptions about the inexorability of historical outcomes and notions of progress, is one of the most important jobs of the historian.

Choices by powerful interests have shaped every megaproject on Earth, but the longer such structures become a permanent part of the landscape, the harder it becomes to envision other conditions of possibility and alternative futures.[46] To examine bygone debates over unbuilt projects reminds us of those multiple options, and of the negotiations required to bring large public works to fruition in democratic societies. For those now seeking to mitigate the threats to human rights and ecological quality posed by development plans, the sea-level canal story offers a hopeful precedent for challenging narratives of inevitability—but also an ominous reminder that environmental management, like nature itself, evolves along multiple, nonlinear paths and not necessarily toward progressive ends.

PART I

In the Shadow of the Panama Canal

CHAPTER 1

Canalizing and Colonizing the Isthmus

FOR OVER A CENTURY authors have celebrated the Panama Canal as a triumphal conquest of nature. Hundreds of books and articles have honored its visionary engineers, the enormous amounts of dirt removed, the acres of wetlands drained to control tropical diseases, and other superlatives that speak to the monumental transformation of the isthmus. But other important parts of the canal story have emerged in recent decades. Influenced by labor history, anthropology, environmental history, and science and technology studies, the newer literature centers the working-class people who built the structure amid intense racial injustices, the communities swept aside to create the canal landscape, the maintenance issues that undermined the conquest-of-nature narrative, the scientific knowledge generated in the context of the altered isthmian environment, and other stories that enable us to see the artificial waterway as something much more than an amazing technological testament.[1]

This chapter explores another aspect of the Panama Canal's history that ebbed from public consciousness as the massive structure took shape—the prolonged controversies over where to site it and how to design it. The isthmian canal question captivated the nineteenth century's most famous scientist, the Prussian naturalist and explorer Alexander von Humboldt (1769–1859). The routes he identified in 1811 as most suitable drew attention not only during his long life. Like a dormant caterpillar awaiting the right external conditions, they reemerged an astounding 150 years later, when the then aging waterway needed updating (maps 1.1 and 1.2).

Over the course of almost five decades, Humboldt experienced frustration in his quest to mobilize resources for comprehensive isthmian surveys, and he changed his mind about where and how the "water communication" should be built. His canal advocacy illustrates the contingent nature of megaproject planning. It also accentuates the environmental challenges and imperial blind spots underlying the long-standing plans for an Atlantic-Pacific link.

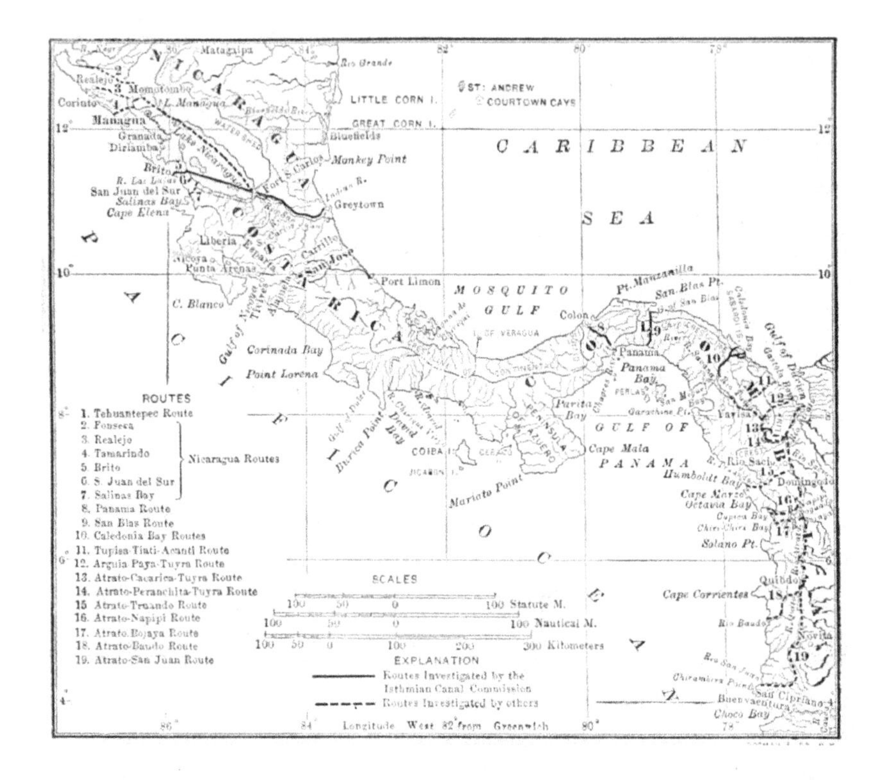

MAP 1.1. A 1902 map of the Central American isthmus demonstrating nineteen possible routes for an interoceanic canal, several of which Humboldt addressed during the first half of the nineteenth century. William Hubert Burr, "The Panama Route for a Ship Canal," *Popular Science Monthly* 61 (1902): 257.

The chapter then examines the post-Humboldtian alignment of political, economic, military, and technological forces in favor of the Panama route and against the sea-level design. This approach helps us appreciate the central Panamanian lock canal as the product of a specific coalition of stakeholders who seized the right opportunities at the right times, rather than as the expected outcome of strategic geography and U.S. technopolitical superiority.

Humboldt and the Changing Canal Calculus

Ever since the Panamanian isthmus became a global route for conveying Peruvian gold and silver to Atlantic ships during the sixteenth century, those searching for a natural maritime passage had pondered the possibility of creating an artificial one. Eventually, the Spanish government developed a road-and

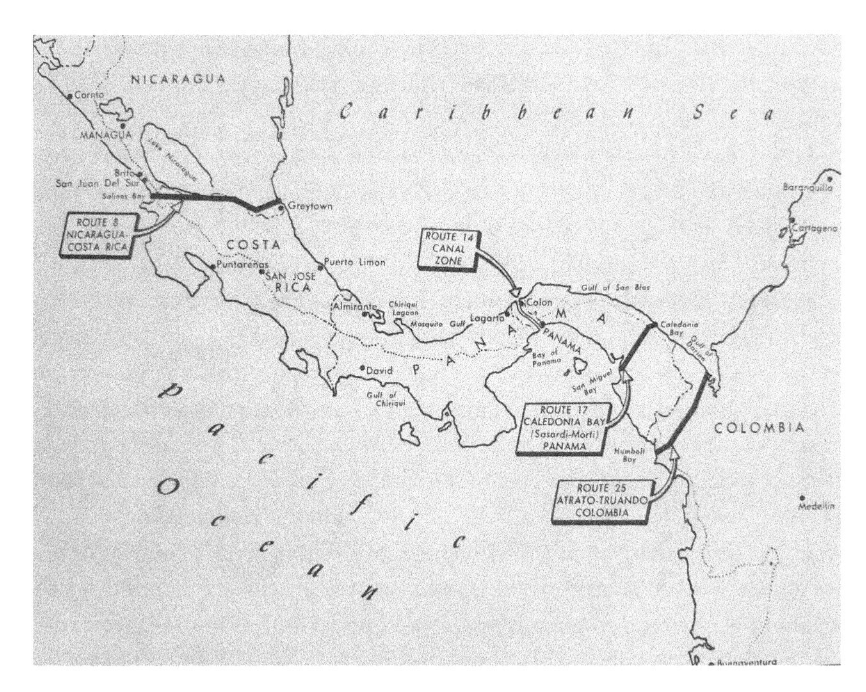

MAP 1.2. The routes investigated by the Atlantic-Pacific Interoceanic Canal Studies Commission. The two main nuclear routes, in eastern Panama and western Colombia (Routes 17 and 25), are to the right. *Atlantic-Pacific Interoceanic Canal Study Commission*, p. 10, Entry A1 36040-D, Container 8, RG 220, U.S. National Archives and Records Administration, College Park, Md.

river-based transportation network that connected the oceans. Not until 1814, on the eve of independence of the Spanish Latin American colonies, did the crown manifest interest in cutting a canal from the Caribbean to the South Sea (as the Pacific Ocean was often called).[2]

Foreign institutions such as the French Academy of Sciences had raised the isthmian canal issue during the Age of Enlightenment, but the person who put the project on the agenda of nineteenth-century world leaders was Humboldt. "The Philosopher" (as the front page of the *New York Times* memorialized him on the one hundredth anniversary of his birth) is well known to historians of science and postcolonial scholars as a pioneering biogeographer and critic of environmental mismanagement by the Spanish Empire.[3] Because artificial waterways have a long history of disrupting ecological and social communities, his lifelong advocacy for a project as damaging as an interoceanic canal seems at odds with his reputation today as a "bracingly contemporary" prophet of the

Anthropocene.[4] In fact, it demonstrates how taken for granted the idea of improving the environment via massive civil engineering works was (and in some contexts, still is).

Humboldt achieved worldwide fame through popular accounts of his travels and studies of geography, geology, astronomy, meteorology, and ecology, among other fields. A large inheritance enabled the thirty-year-old mining inspector to finance his own scientific expedition to the Americas from 1799–1804, for which he convinced the king of Spain to issue him and his partner rare passports. Although he did not visit any parts of the isthmus south of modern-day Mexico, Humboldt's access to maps, letters, and engineering reports in the archives of the Mexican viceroyalty and other Spanish territories provided crucial information about potential routes for a large artificial waterway.[5] He discussed the issue in three best-selling books spanning four decades: *Political Essay on the Kingdom of New Spain* (1811); *Personal Narrative of Travels to the Equinoctial Regions of America* (1826), which contained the most detailed analysis; and *Views of Nature* (1849). Until his death in 1859, Humboldt endeavored to persuade the European and Latin American powers to overcome what he deemed their baseless concerns and invest in "a communication between two seas, capable of producing a revolution in the commercial world."[6]

Which part of the New World would best accommodate a ship channel uniting the Atlantic and Pacific Oceans? In the *Political Essay*, Humboldt described nine routes or, as his translator put it, "points" for cutting a canal. Five spanned the Latin American isthmus, at Tehuantepec (Mexico), Nicaragua, central Panama (which he broke down further into three pathways), and two sites in northwestern Colombia designed to utilize the Atrato River, which flows north into the Caribbean Gulf of Darién (one route connected to the Pacific Cupica Bay and the other to a rumored artificial waterway further south known as the lost Raspadura Canal).[7]

Humboldt lamented the failure of previous explorers to apply a rigorous scientific approach to the physical geography of these regions. In particular, although the fifty-mile route from the Caribbean to the Bay of Panama had "occupied every mind" since Vasco Núñez de Balboa's crossing in 1513, fundamental questions remained about the elevation of the cordilleras and whether the oceans on either side were of different levels. As he exclaimed in his interdisciplinary way, "These are problems whose solution is equally interesting to the statesman and the geographical naturalist!"[8]

Humboldt revisited the issue in the sixth volume of his *Personal Narrative*, for which an English translation appeared in 1826. By that time, most of the Spanish colonies in the Americas had gained emancipation, and the governments of the newly independent nations spanning the isthmus were beginning to explore the prospects of canal construction.[9] In addition, the state of New York in the northeastern United States had completed in just six years a 363-mile-long barge canal between Lake Erie and the Hudson River. The Erie Canal connected the Atlantic port of New York City with the upper Great Lakes, and rapidly recouped its cost. For Humboldt, upstate New York's artificial river provided an impressive example of the ability to open up trade and overcome the enormous expenses of excavating mountainous terrain.[10]

Despite his dismay that statesmen and merchants still lacked the geodetic data needed to make the right choice about the isthmian waterway's location and magnitude, Humboldt did not let the lack of evidence stop him from stating his opinions.[11] Of the five routes, he declared, "The isthmus of Nicaragua and that of Cupica have always appeared to me the most favourable for the formation of *canals of large dimensions*, similar to the Caledonian canal," the fifteen-foot-deep waterway across central Scotland linking the Atlantic with the North Sea.[12] He had long since ruled out the central Panama routes on the erroneous assumption that the mountains there were too high for ditchdigging.[13] He also now asserted—incorrectly as he found out toward the end of his life—that the mountain range between the Atrato River and Cupica Bay lowered to such a degree as to disappear. Despite the apparent topographical appeal of the Colombian Darién route, however, Humboldt conceded the primacy of geopolitics: "It appears somewhat probable that the province of Nicaragua will be fixed upon for the great work of the junction of the two Oceans."[14] Nicaragua's proximity to the United States did later make it very attractive to investors in the north.

Humboldt called for the new Latin American nations to fund engineering surveys of each of the five major routes so as to make an informed decision, and thereby persuade "governments and enlightened citizens" to buy shares in a joint-stock company to finance a transisthmian water communication.[15] He warned that the construction process would present unprecedented challenges to the Old World's hydraulic experts: "The facility of collecting an enormous mass of rain waters within the tropics [for feeding a canal], is beyond what the engineers of Europe can imagine." Because the tropical forests received at least five times as much rain as Paris, the canal designers would have to take many more variables into account than, say, the landscape architects of Versailles's famous irrigated gardens.[16]

Despite his reputation today as a founder of modern environmental thought, Humboldt did *not* address the issue that would most interest modern environmentalists: the destructive ecological effects of deforesting a large swath of the Continental Divide, removing one hundred million cubic meters of earth, and using the material to form new dams and causeways. That is not to say that deforestation and other transformative human activities did not concern him; on the contrary, he published pioneering critiques of the damaging unintended consequences of clearing tropical lands for plantation agriculture and urban development.[17] Yet having worked in the mining industry, Humboldt, like all civil engineers, sought to improve the natural environment for human use and convenience.[18]

For Euro-American captains of industry, what could be more convenient than a shortcut linking the Pacific and Atlantic realms? An isthmian waterway would revolutionize world trade by precluding the long voyages around Cape Horn and the Cape of Good Hope. Moreover, it would radically alter East Asian relations with western Europe and North America; in Humboldt's words, "That neck of land against which the equinoxial current breaks, has been for ages the bulwark of the independence of China and Japan."[19] Subsequent events, such as the Anglo-Chinese Opium Wars and U.S. commodore Matthew Perry's 1853 expedition to Japan, undercut the patronizing idea of East Asian independence as a function of inefficient European access. Nevertheless, framing the Central American waterway as an agent of globalization and Asia-Pacific transformation was prescient.

The only concern Humboldt conveyed in public regarding the canal was the potential for military conflict. He foresaw the possibility that powerful nations might wage war to control the conduit, confessing, "I am not secured from that apprehension either by my confidence in the moderation of monarchical or of republican governments, or by the hope, somewhat shaken, of the progress of knowledge, and the just appreciation of human interests."[20] Indeed, his compadre Simón Bolívar, the revolutionary leader who in 1819 became president of Gran Colombia (a nation encompassing present-day Colombia, Panama, Venezuela, and Ecuador), had rejected an application for a concession to build a canal in 1821 for fear that it "might afford facilities to the enemy" for recolonizing Latin America.[21] Such events gave Humboldt reason to doubt that progress and allied Enlightenment values could be sustained far into the future—let alone in the present so as to overcome the epistemic, technological, economic, and political obstacles to solving the canal problem.

Humboldt returned yet again to the interoceanic transit issue in 1849's *Views of Nature*. The discovery of gold in California had caused westward traffic to explode around Cape Horn and across Central America, especially in Panama.

The difficulty of crossing the Panamanian landmass, though only fifty miles long, by canoe and mule reignited interest in more efficient forms of transportation infrastructure. Yet geographic data remained scarce. Granted, General Bolívar had long since granted Humboldt's request to commission a survey between Panama City (on the Pacific coast) and the mouth of the Chagres River (on the Caribbean/Atlantic coast).[22] It had led to other investigations of central Panama, yet as Humboldt exclaimed, "The most important points on both the eastern and southeastern portions of the isthmus on both coasts have been ignored!" He reprised his call for precise topographical determinations of the entire isthmus, especially the southeastern portion "where it connects to the mainland of South America at the Darién Gap."[23] Yet he omitted the Nicaragua route from the 1849 discussion, a revisionist approach that was telling of the contingency of the canal problem.

In the last decade of his life, Humboldt's hope for a Darién survey seemed on the verge of fulfillment. In 1854, fearing competition from British and French interests, the U.S. Navy launched the first Darién Exploring Expedition. The party planned to investigate the rumored forty-mile route from the Atlantic Caledonia Bay to the Gulf of San Miguel, an area north of the Cupica Bay route that Humboldt had appeared to endorse in an 1853 letter.[24] However, malaria, madness, and starvation cut short the scientific reconnaissance. Popular accounts of the ninety-seven-day ordeal reified perceptions of the Darién as a dangerous wilderness inhabited only by remnants of Indigenous Guna who had survived Spain's genocidal wars. Tropical diseases and famine had also doomed an infamous 1698 colonization effort that bankrupted Scotland. A century later, even the Spanish retreated from the dense forests and swamplands of the ten-thousand-square-mile mountain pass.[25]

Before, during, and immediately after the expedition's mortifying failure, U.S. corporations concentrated on developing new land-based transportation networks to carry California-bound travelers across the isthmus. Central Panama became the site of the first transcontinental railroad in 1855.[26] Despite its success, U.S., British, and French teams conducted private and state-supported canal surveys across the nations of Central America during the subsequent decades.[27]

Seven expeditions received funds from a Wall Street financier, sea-level canal enthusiast, and Darién canal concession holder named Frederick Kelley. Kelley credited the writings of "the illustrious Humboldt" and Admiral Robert Fitz-Roy, who had captained the famed *Beagle* voyage of Humboldt's disciple Charles Darwin, with sparking his interest in the Darién routes.[28] In 1856, Humboldt validated Kelley's quest by receiving him at his home in Berlin and writing him a letter that several outlets reprinted.[29] In earlier publications, Humboldt had not said whether the ship channel should be at sea level, but he now came out on Kelley's side: "The great object to be accomplished is, in my opinion, a canal uniting the two oceans *without locks or tunnels*."[30] However, Kelley lost his concession and fortune before he could fulfill his hero's dream. Not until the 1870s did surveyors return to the southeastern Darién sites that had so interested the illustrious naturalist in his final years.[31]

The Contingent World Wonder

In 1869, a decade after Humboldt's death and the year that citizens of both North and South America celebrated the centennial of his birth, the French diplomat and entrepreneur Ferdinand de Lesseps oversaw the completion of a technological sensation. The 120-mile-long Suez Canal joined the Mediterranean and Red Seas, cutting 4,300 miles off the voyage between the North Atlantic and northern Indian Ocean. Funded by the sale of shares in the Suez Canal Company, the project took ten years and the lives of thousands of workers, many of them enslaved. Yet the transformation of the desert isthmus into a moneymaking maritime highway cemented France's reputation for cutting-edge civil engineering, and intensified interest in creating a similar bypass between the Pacific and Atlantic.[32]

Ulysses S. Grant made the isthmian waterway the subject of his first address to Congress after assuming the U.S. presidency in March 1869, and during his tenure, seven state-sponsored expeditions conducted surveys that built on and refined Humboldt's routes. For example, as one surveyor, Lieutenant Frederick Collins, tactfully noted, Humboldt was "somewhat misled" as to the actual height of the mountain range near Cupica Bay, the Pacific terminus of one of the Darién sites. Collins argued in 1874 that enough data had been collected to narrow the choice down to three possible routes, none of which included central Panama: "We need consider only Tehuantepec, Nicaragua and the Napipi-Doguado [two river valleys linked by the Atrato], for at one of these three points the canal will surely be built, if built at all." He then dismissed the Mexican and Nicaraguan options due to "the earthquake question," making for only one practicable choice, despite the downside of having to tunnel through Colombia's mountainous terrain.[33]

Yet despite the historical record of earthquakes in Nicaragua, the route held significant advantages for U.S. interests. It was closer to New Orleans and other U.S. ports than the Panamanian or Colombian sites and easier to excavate than the Tehuantepec route due to the presence of the 103-mile-long Lake Nicaragua and the Caribbean-flowing San Juan River. On the other hand, the lake's high elevation would require locks to lift and lower ships by as much as 110 feet. Grant appointed a commission to review the conflicting conclusions of the various isthmian expeditions, and in 1876, the three members deemed the Nicaragua route the most advantageous "from engineering, commercial, and economic points of view."[34]

But the French beat the Americans to the punch. Shareholders in the Suez Canal Company had experienced handsome returns during its first decade, and buoyed by de Lesseps's interest in replicating his success in Central America, thousands of French citizens bought shares to build a sea-level canal in the one place Humboldt had firmly rejected: central Panama, roughly parallel to the railroad. Although the mountains there were only a third as high as Humboldt had thought, the heavy rains he had warned of magnified the tendency of the clay-streaked soil to collapse on itself. Even worse than the downpours and landslides were the horrific outbreaks of malaria and yellow fever, which killed twenty-five thousand workers. Carving a channel through rainforests and wetlands did not compare to digging in the Egyptian desert, and by 1889, de Lesseps's project and career imploded.[35]

Even as another French company tried to salvage the project by resuming work on a smaller scale from 1894 to 1904, U.S. canal fever remained strong. The 1890 publication of a book by a Naval War College history professor provided a justification for an Atlantic-Pacific link that transcended Humboldt's focus on commercial exchange. Alfred Thayer Mahan's tome *The Influence of Sea Power upon History, 1660–1783* inspired politicians chastened by the nation's economic downturn and by perceptions of the frontier's closure. Rather than accepting that the era of Manifest Destiny was over now that white settlers had filled the lands west of the Mississippi, Mahan argued that Americans must extend their military dominion over the oceans to protect and expand their commercial fortunes, just as Great Britain had become a world power via naval supremacy, maritime trade, and a far-flung colonial network. Mahan called for developing naval bases outside U.S. boundaries. Like islands providing temporary habitat for land birds unable to fly far offshore, such structures would sustain the nation's fleet of battleships: "To provide resting-places for them, where they can coal and repair, would be one of the first duties of a government proposing to itself the development of the power of the nation at sea."[36]

Securing coaling stations throughout the Caribbean Sea would be especially important "if a Panama canal-route ever be completed." Such a connection would transform the region into a great commercial highway, attracting throngs of European ships and precluding the United States from "stand[ing] aloof from international complications." Accordingly, no matter who controlled the isthmian canal, U.S. naval ships would need to patrol the Caribbean to defend the southern borders and to project hemispheric authority: "With ingress and egress from the Mississippi sufficiently protected, with such outposts in her hands, and with the communications between them and the home base secured . . . the preponderance of the United States on this field follows, from her geographical position and her power, with mathematical certainty."[37]

Mahan's arguments tapped into a broader pool of interest in modernizing the U.S. Navy and facilitating a paradigm shift toward sea power.[38] The benefits of investing in battleships as instruments of economic and foreign policy paid off enormously for the United States in the Spanish-Cuban-American War of 1898, after which Spain ceded Cuba, Puerto Rico, Guam, and the Philippines.[39] As Mahan wrote years later, "From a military point of view, these acquisitions have advanced the southern maritime frontier of this country."[40] Defending the new U.S. empire's tropical frontier, which also included Hawaii, necessitated a Pacific-Caribbean passage more than ever. This was epitomized during the war by the dramatic voyage of the U.S.S. *Oregon*, which took sixty-six days to rush from San Francisco to the Cuban battlefront by way of Cape Horn; as many observers noted, an isthmian shortcut would have shaved off eight thousand miles.

Several political, economic, and technological forces converged in the half decade from 1898 to 1903 to make the U.S.-controlled Panama Canal a reality. Underlying them all was Theodore Roosevelt's fervid advocacy. As assistant secretary of the navy during the war, he had helped implement Mahanian goals, and after becoming president in 1901, he lobbied hard for a canal in Panama. Although the House of Representatives had voted for a bill specifying a Nicaraguan route, he and a few key stakeholders seeking to build on the French project convinced the Senate otherwise. Part of their infamous strategy took advantage of a vintage Nicaraguan stamp featuring a spewing volcano, along with the recent news of an eruption elsewhere in the Caribbean, to undermine confidence in the route.[41] Congress passed a law authorizing the president to purchase the French syndicate's assets and rights and to build an isthmian canal, contingent on the negotiation of a treaty with Nicaragua or Colombia, of which Panama had long been a province.

When the Colombian Senate objected to the ensuing U.S.-Colombia treaty on financial grounds, well-connected insiders encouraged dissatisfied Panamanian elites to revolt in 1903. Roosevelt deployed battleships to the Caribbean and Pacific coasts to prevent Colombian troops from suppressing the revolution. No Panamanians took part in the ensuing treaty negotiation, which was led by a French canal agent who benefited from the sale of his company's assets for $40 million. Phillippe Bunau-Varilla helped craft the 1903 treaty that bore his name and granted the United States extremely favorable terms: the right to build and defend a canal and surrounding zone over which it could rule as "if it were the sovereign" in exchange for a $10 million payment and $250,000 annuity. The inequitable treaty ignited resentment among Panamanians from its inception.[42]

Having secured, in essence, sovereign rights over the Panama Canal Zone, U.S. officials postponed the decision over whether to continue with de Lesseps's plans for a sea-level channel. Higher priorities faced the engineers: getting the aged railroad in shape to carry away dredge spoil as fast as the huge steam shovels could dig, and taming deadly tropical disease organisms, an enormous task that drew on twenty years of epidemiological research.[43]

Roosevelt appointed a board of consulting engineers in 1905 to settle the design issue. He told the thirteen members that he hoped it would be feasible to excavate the entire route at the level of the seas: "Such a canal would undoubtedly be best in the end . . . and I feel that one of the chief advantages of the Panama route is that ultimately a sea-level canal will be a possibility." However, facilitating interoceanic traffic as soon as possible outranked "the ideal perfectibility of the scheme from an engineer's standpoint." A sea-level waterway would shorten transit times, but Roosevelt did not consider it worth adding too many years and safety risks to the construction process. Still, if the board recommended a high-level multilock canal as the most expeditious plan, he desired to know whether it could be converted to sea level at some future point "without interrupting the traffic upon it."[44]

The board voted eight to five for the sea-level canal option as the only one "giving reasonable assurance of safe and uninterrupted navigation." Other lock canals caused "vexatious delays" and accidental collisions, and provided ideal targets for violent adversaries: "The modern lock for ocean-going vessels is a work which an enemy, through stratagem, could with no great difficulty put out of use in an hour or even a few minutes." By sneaking in dynamite via the surrounding forest or blowing up a transiting ship, saboteurs could disable the canal for months. Moreover, while recognizing that a seaway would cost more and take longer to build, the majority stressed that it would "endure for all time" no matter how large ships might become, and that the construction expenditures (estimated at $250 million and twelve to thirteen years) must be balanced against those of maintaining and defending a vulnerable locked waterway.[45]

The dissenting board members, however, argued that their colleagues minimized the difficulties posed by the eight-mile-long Culebra mountain portion of the Continental Divide, known as the Culebra Cut. De Lesseps's crews had managed to lower the summit from 210 feet to 193.5 feet above sea level. Excavating all the way down, even using the latest technology, would take closer to fifteen years—six years longer than a lock waterway with a summit elevation of 85 feet, or as they called it, a summit-level canal. It would also cost $100 million more, "not a trifling sum, even for the resources of the United States."[46]

The minority report engineers also rejected the majority's assessment of the disproportionate risks posed by lock canals. Even a sea-level canal, they argued, would be vulnerable to attack at its tidal-regulating structure, a device needed to account for the different maximum tidal ranges on either side of the isthmus. As surveyors had quantified several decades previously, the Pacific tides rise and fall twenty feet each day, whereas the Caribbean tides move only two feet. In a sea-level channel, depending on the time of day, the resulting current flows might compromise navigational safety. A tidal-regulating structure at the Pacific terminus would enable engineers to adjust the currents at a given moment to the desired velocity—but of course not if saboteurs took them out.[47]

As for the question of accommodating likely increases in ship size, the pro-lock group questioned the wisdom of trying "to meet the possible requirements of a distant future, which might be estimated erroneously and would burden the commerce of the present and near future with unfavorable conditions."[48] Larger locks could be built if and when needed. Though Roosevelt's question about the mechanics of conversion remained unanswered, he could not have agreed more with the board's minority members.[49] Construction on the summit-level canal proceeded in 1906, and within a few years, the idea that there had been an alternative receded from public memory: "The controversy that once raged so furiously now seems to have been but a tiny tempest in an insignificant teapot."[50]

Girdling Panama

The megaproject imposed immense changes on the Panamanian landscape. To control the insect-borne diseases that had killed so many of de Lesseps's workers, sanitarians drained wide swaths of wetlands, installed drinking-water systems in the port cities of Colón and Panamá, and sprayed hundreds of thousands of gallons of oil and larvicide.[51] To carve out the canal bed, workers excavated over 150 million cubic meters of rock and soil—enough to create both the two-mile-long Amador Causeway guarding the waterway's Pacific entrance and the world's largest earthen dam. Closer to the Atlantic side, the 1.5-mile-long Gatun Dam channeled the Chagres River into the world's largest reservoir, Gatun Lake, to serve as the canal's water and electricity source. To move so much soil, crews from Barbados, Jamaica, and many other nations mobilized massive steam shovels and hydraulic rock crushers shipped from the booming factories of the Ohio River Valley and Great Lakes. They also installed three pairs of Pittsburgh-forged locks—each measuring 110 feet wide by 1,000 feet long—to lift ships 85 feet above sea level. The ingenious "bridge of water" used fifty-two million gallons of fresh water from the Chagres for every transit.[52]

U.S. officials also disrupted the region's human communities by depopulating towns to make way for the ten-mile-wide Panama Canal Zone, an enclave designed to house the waterway's civilian employees and military defenders (map 1.3). The radical reshaping of central Panama's human-dominated landscape was not an unavoidable byproduct of canal construction. Rather, it resulted from specific decisions that benefited the United States—choices that were easy to forget as second-growth forests took root in cleared fields, and as the waters of Gatun Lake submerged what had been for over three centuries an intensively cultivated valley.[53]

Creating the massive bridge of water and its buffer zone required technological, scientific, and organizational expertise, which countless magazine spreads, postcards, popular books, and world's fair exhibits commemorated with jingoistic flair.[54] "Every American can take a just pride in this girdle which we have flung across the isthmus," enthused one author, especially since "we are the nation which . . . Providence . . . has decreed should build the canal . . . to confer a lasting benefit on the world at large and usher in a new age of culture."[55] Americans also took pride in photographs depicting Roosevelt, who defended his actions against Colombia, visiting the construction site. The image of him operating a steam shovel in a white linen suit became a powerful icon of the conquest of nature and other nations. Grade school U.S. history textbooks replicated such heroic representations throughout the twentieth century, reinforcing belief in the project's inexorableness and righteousness among generations of U.S. citizens.[56]

Publicity regarding the Panama Canal played up its international commercial benefits and the munificent U.S. policy of keeping tolls low rather than trying to recoup the $400 million cost. Of course, as later analyses revealed, low tolls functioned as a subsidy for U.S. shippers moving goods from coast to coast.[57] The elderly Mahan, not surprisingly, stressed the incalculable national security benefit of moving the U.S. naval battalion between oceans as needed. While

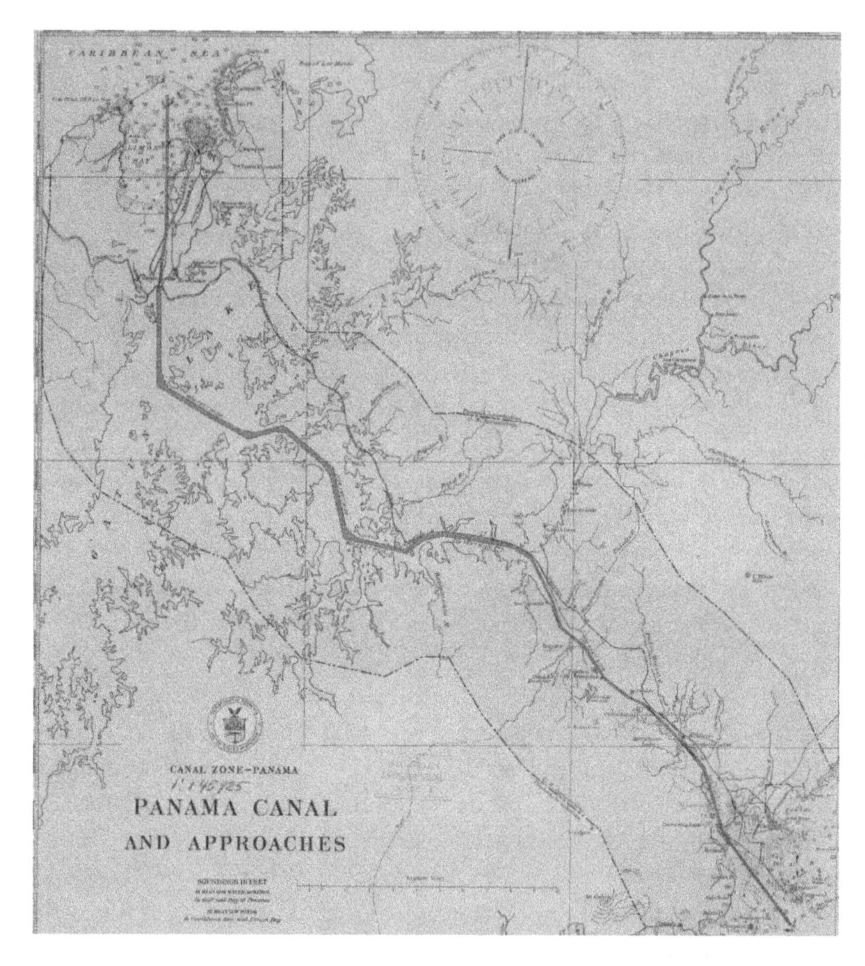

MAP 1.3. A 1914 U.S. Coast and Geodetic Survey map of the Panama Canal Zone, which extended 5 miles from each side of the waterway and covered 550 square miles. NOAA Central Library, Silver Spring, Md.

the fleet would have to be maintained in the Atlantic for the foreseeable future due to the West Coast's inferior coal deposits and high labor costs, ships could steam from Norfolk, Virginia, to Pearl Harbor, Hawaii, in four weeks rather than four months.[58]

Perhaps less expectedly, the retired naval historian-officer also promoted the waterway as a bastion against what Roosevelt and other nativist contemporaries called "race suicide."[59] Viewing the West Coast as underpopulated and in need of more white immigrants, Mahan declared, "The great effect of the Panama

Canal will be the indefinite strengthening of Anglo-Saxon institutions upon the northeast shores of the Pacific, from Alaska to Mexico, by increase of inhabitants and consequent increases of shipping and commerce." Passenger ships transiting the new canal would enable white Europeans and East Coast residents to make the journey at a lower cost than the transcontinental railroad or Great Lakes steamers.[60]

Conclusion

How would Humboldt have responded to such martial and white supremacist rationales for the transisthmian canal? Probably not favorably. He had expressed explicit concern about nations fighting to control such a conduit, and more broadly, he rejected scientific racism and its allied institution, slavery. "Whilst we maintain the unity of the human species," he wrote in the blockbuster first volume of *Cosmos* in 1845, "we at the same time repel the depressing assumption of superior and inferior races of men."[61] Moreover, his critiques of Spanish colonial policies that degraded human and ecological communities had bolstered the Latin American independence movement led by Simón Bolívar.

At the same time, Humboldt's canal advocacy must be seen in the context of his contested role as an agent of imperialism.[62] Speculation that he shared his canal intelligence prior to publication with U.S. president Thomas Jefferson supports the view of him as a proponent of using scientific knowledge and tools to promote Euro-American dominance and Northern Manifest Destiny.[63] Due to the problematic imperial as well as environmental dimensions of the canal enterprise, more historiographical attention to his private and public writings on the subject might help address the question of whether Humboldt deserves his reputation as the founder of modern environmental thought.[64]

For the purposes of this book, Humboldt's English language publications in favor of the canal illustrate the historical contingencies of megaproject planning. Despite being the world's most famous scientist—one who succeeded in getting other projects off the ground (such as networks of magnetic and meteorological observatories)—he failed for forty years to convince officials to conduct comprehensive surveys of all the routes he had identified. Conducting the scientific reconnaissance work, let alone the large-scale engineering of the actual structure, required favorable political, economic, and technoscientific forces to coalesce at the right moments.

Humboldt's advocacy also challenges notions of geographical and historical determinism that permeate popular writings on the Panama Canal.[65] A determinist perspective emphasizes notions of inevitability and predestination. Consider this quote, crafted the year after the maritime highway opened for business: "The valley of the Chagres was framed by the hand of Nature in such a way as to fit admirably into the plans of the canal engineers for a lock canal across the isthmus, with the Atlantic locks at Gatun."[66] Of course, for Humboldt and other nineteenth-century canal enthusiasts, it was not obvious that a lock design would prevail nor that the Chagres River valley in central Panama offered the ideal site. As a U.S. senator wrote in 1837 of the southern Atrato River valley, "Nature seems to have designed this for the passage. The Andes are here for a moment lost, and in obedience to the will of Providence and the wants of man, seem to have defiled [narrowed], that commerce may march from the old world to the new."[67]

By the time of the Panama Canal's completion, people had already begun to forget the alternatives that had been the subject of intense debate for decades. Popular authors depicted de Lesseps's failure as a foil to the U.S. initiative, conceding that the French provided "the knowledge that made it possible for us to avoid their mistakes and profit by their experience."[68] Knowledge of the tangible and intangible things needed for a sea-level canal was indeed valuable and worth remembering, as events would prove sooner than expected.

CHAPTER 2

Confronting the Canal's Obsolescence

THE PANAMA CANAL'S imposing physical footprint, and the propensity of its popular chroniclers to end the story in 1914, has obscured an important fact: within a generation, its owners feared it was becoming obsolete. In fact, the agency responsible for operating and maintaining the waterway, the Panama Canal Company (PCC), initiated an ambitious expansion in 1939. But the project stalled in the midst of World War II, and by the time the fighting ended, larger locks no longer appeared a worthwhile investment. Aerial warfare could disable them overnight. For officials seeking to bomb-proof the canal at the start of the Atomic Age, dusting off old blueprints for a low-tech seaway made more sense than enlarging vulnerable locks.

Decision makers balked when presented with the exorbitant quotes for converting the Panama Canal to sea level or excavating a new channel along one of the routes studied in the previous century. But in the 1950s, an exciting new technology offered a cheaper, safer, and more efficient means of moving massive quantities of earth: peaceful nuclear explosives (PNEs). Rather than weaponizing atomic energy, PNE designers worked to harness it to reshape the environment for the use and convenience of humankind. At last, civil engineers would be liberated from the crushing constraints posed by "the physics of the isthmus."[1]

Resurrecting old sea-level canal plans in the context of PNEs was not only about meeting the needs of modern shipping and military transport. Peaceful nuclear excavation boded well for resolving two kinds of obsolescence facing the postwar Panama Canal, the technological and the geopolitical.

The 1903 Hay–Bunau-Varilla Treaty had granted the United States the perpetual right to use the lands surrounding the canal as "if it were the sovereign." The resulting 553-square-mile Canal Zone angered Panamanians by functioning as a colony in all but name. It bisected the entire nation, hosted several U.S. military bases, and housed the canal's white "gold roll" civilian employees in manicured suburbs that were nothing like the neighborhoods designated for nonwhites on the "silver roll."[2]

After the Suez Crisis and the Cuban Revolution, the Cold War competition with the Soviet Union to win hearts and minds in the so-called Third World focused attention on the inequities perpetuated by the United States in Panama. Although previous anti-Yankee protests had led to a few concessions, violent demonstrations in 1947, 1958, 1959, and, most significantly, 1964 attested to the unsustainability of the status quo.

But how could Washington meet Panamanian demands for treaty reform without sacrificing commercial and hemispheric power? By negotiating new agreements that separated the issues of military bases and the canal, and by prioritizing the development of a simplified waterway that would require few people to operate and defend—a sea-level canal. And what if Panama wanted more? Pitching a cheap nuclear-excavated seaway to another interested client, such as Colombia or Nicaragua, would provide a potent means of checking Panamanian resistance.

This chapter brings the first twentieth-century phase of the sea-level canal proposal out of the historiographic shadows by examining the technological and diplomatic problems the first two generations of Panama Canal officials faced. The postwar modernization studies highlight the roles that the nuclear seaway played for Presidents Dwight D. Eisenhower and John F. Kennedy as they sought to blunt the sharp edges of U.S. imperialism while maintaining control of a strategic region.

Obsolescing Shibboleths

Even as popular writers extolled the almost-completed Panama Canal as a wonder of the modern world, serious challenges remained to be solved. As one author conceded in 1913, "Of course, a vast amount remains to be done, and capricious nature may devolve extra labor upon us if she persists in trying to close the cut we have so laboriously excavated at Culebra."[3] In fact, landslides at the Culebra Cut continued to cause dangerous obstructions in the decades following the canal's opening in 1914, necessitating frequent, expensive rounds of dredging (fig. 2.1).[4]

Another threat posed by capricious nature involved disruptive variations of rainfall entering the Chagres River watershed. Despite the immensity of the canal's water source, the 163.5-square-mile Gatun Lake, it could not store enough water during droughts for the increasing traffic. Nor could the Gatun Dam control all upstream deluges, forcing occasional canal closures. U.S. officials responded by invoking the clause of the Hay–Bunau-Varilla Treaty that allowed them to expropriate land from Panama for canal maintenance and protection, and built the Madden Dam on the mountainous upper Chagres from 1931 to

FIGURE 2.1. A severe landslide blocking the Panama Canal, 1916. Sea-level canal advocates argued that their design would preclude dangerous blockages caused by landslides, ship collisions, or acts of sabotage. Prints and Photographs Division, Library of Congress, LC-USZ62-55958.

1935. The resulting 22-square-mile Alajuela lake provided an additional freshwater supply for operating the locks and generating hydroelectricity. It also flooded out more forests and people and expanded the dominion of Canal Zone officials.[5]

A third major problem was not as amenable to a quick technological fix: naval and cargo vessels were growing in size, just as sea-level canal proponents had predicted. In 1929, Congress authorized feasibility studies to determine the cost of expanding the existing waterway and of building a second one in Nicaragua, with which the United States had signed a canal treaty in 1914.[6] The colossal expenses for both endeavors led officials to concentrate instead on the Madden Dam project, but pressure coalesced in the late 1930s in favor of expanding the existing waterway. Congress allocated $277 million for a third lane of locks, each measuring 1,200 feet long, 140 feet wide, and 45 feet deep. As a PCC publication announced in June 1941, shortly after excavation began on the New Gatun Locks, "The original builders were aware that the Canal's capacity would eventually require enlargement," but not so soon.[7]

Construction stopped in 1942, however, and officials focused instead on preparing for a Pearl Harbor–style attack by stocking extra lock gates and building new defensive infrastructure within and outside the Zone.[8] After the war, the atom bomb cast doubt on the viability of lock canals. The attacks against Hiroshima and Nagasaki, and the experimental detonations in the South Pacific, showed the staggering effect of exploding the equivalent of thousands of pounds of TNT: "One has only to recall the movies of the Bikini atomic bomb tests," wrote one engineer in 1947, "where a column of water that looked to be half a mile across was thrown several thousand feet in the air, to visualize what would happen to a lock."[9]

Such sobering thoughts resurrected the "sea-level canal ghost."[10] At the 1946 annual meeting of the American Society of Civil Engineers, one of the Panama Canal's longest-serving employees invoked the 1906 consulting board's majority decision about the invulnerability of sea-level waterways. He called for converting the famous bridge of water to a sea-level channel, as did others seeking a "bomb-proof canal."[11]

Congress authorized the president of the PCC (who also served as governor of the Canal Zone) to initiate a new modernization study in December 1945. PCC personnel worked closely with the Army Corps of Engineers, the venerable federal agency that had played a key role in building the Panama Canal, to complete the report, titled *Isthmian Canal Studies—1947*. The investigation included a meta-analysis of historical surveys of thirty possible routes spanning Central America, eight of which now appeared suited to sea-level excavation. The report also drew on new geological mapping and exploratory drilling work, and even featured the construction of a half-mile-long hydraulic model that engineers used to measure the effects of currents and tidal-regulating structures. Finally, the authors consulted with experts in soil mechanics, dynamics, and seismology to determine the effects of nuclear bombing on canal structures.[12]

In place of the triumphant portrait that had prevailed since the start of the century, the 1947 engineering report painted a grim picture of the Panama Canal. A "determined and resourceful enemy" could shut it down at any moment. Once breached, the Gatun Dam would empty into the Caribbean, rendering the waterway unusable for one to two years and leaving operators at the mercy of the rains for refilling. If hit with a nuclear weapon, another two years at least would be needed for the radioactivity to dissipate. A generation after it had opened for business, the world wonder now inspired fear rather than confidence: "No lock canal can meet fully the future needs of national defense."[13]

By contrast, a sea-level channel of sixty feet deep and six hundred feet wide would more than justify the $2.483 billion price tag. Granted, it would require special devices to regulate the tidal currents caused by the differences between the tides in the Pacific and Atlantic. Flood-control dams and spillways would also be needed to control river inflows. Nevertheless, the loss of such auxiliary structures would not close the canal for too long. Even an atomic attack would shut it down for a matter of weeks rather than years.[14]

Converting the existing canal to sea level would take ten years and the removal of one billion cubic yards of soil and rock using conventional construction equipment. Also, although ten thousand acres downstream of Madden Dam would be inundated, some of the lands submerged beneath Gatun Lake would resurface, thereby permitting their "return to Panamanian jurisdiction." For all these reasons, the PCC report concluded that conversion constituted "the best means of increasing the capacity and security of the Panama Canal to meet the future needs of interoceanic commerce and national defense."[15]

But the PCC's $2.5 billion solution gained little traction. President Harry Truman submitted the report without comment to Congress, and the ensuing hearings privileged the testimony of atomic warfare specialists that no canal could be rendered bombproof.[16] Looking back several years later, one of the most ardent guardians of the Panama Canal Zone, Representative Daniel Flood of Pennsylvania, denounced the report as "heedless of the diplomatic consequences and costs involved" and praised his colleagues for having "exposed the fallacies upon which it was founded."[17]

Yet the technological status quo could not persist indefinitely, as the Panama Canal's strategic and commercial value slipped throughout the 1950s. The development of air power and a two-ocean fleet had undermined the Mahanian rationale for an Atlantic-Pacific link, and the navy had begun building aircraft carriers that exceeded the dimensions of the locks and the narrow Culebra Cut. Jet planes, the interstate highway, and dieselized railroads could now transport goods and people across the continental United States more quickly than ever. And the voracious industrial expansion of California precluded the need to continue exporting its petroleum and other natural resources eastward.[18]

FIGURE 2.2. U.S. Air Force personnel interacting with Panamanians in Río Salud, Colón, Panama, 1952. The presence of thousands of U.S. civilians and soldiers in the Canal Zone benefited some sectors of the Panamanian economy but also sparked violent anti-U.S. protests in 1947, 1958, 1959, and 1964. Alexander Wetmore, photographer. Smithsonian Institution Archives, Image SIA2009-0018.

Had U.S. control of the Panama Canal become "an obsolete shibboleth?" That was how two Stanford political scientists titled a provocative article in a 1959 issue of *Foreign Affairs* magazine that called for internationalizing the waterway under the United Nations. Not only were military and civilian transportation alternatives undermining U.S. interests in the canal, but so was the increasing "toll in terms of Latin American resentment" exacted by the Canal Zone.[19]

The Zone's fans celebrated it as a beautifully landscaped enclave of tropical suburbs, whose U.S. residents enjoyed subsidized housing, health care, and other perks provided by Uncle Sam.[20] But for Panamanians, the Zone featured segregated towns, U.S. government–run businesses that undercut local entrepreneurship, American flags and military checkpoints, and other daily reminders of Panama's subservient status (fig. 2.2). U.S. officials made some concessions via supplemental treaties in 1936 and 1955. The latter Eisenhower-Remón Treaty increased the annuity to Panama to $1.93 million, transferred some zone lands, provided for a high bridge to be built across the canal, and guaranteed equal pay for Panamanian and U.S. employees in the Zone. Congress delayed implementing the changes, however, and the Hay–Bunau-Varilla Treaty provisions granting U.S. control over the Zone remained in place.[21]

A momentous event in July 1956 drew unexpected attention to the Zone's quasi-colonial status. The Egyptian seizure and nationalization of the Suez Canal intensified U.S. concerns about maintaining control in Panama, leading the Eisenhower administration to revisit the seaway proposal as a technopolitical lever. As the U.S. ambassador in Panama wrote to a State Department official in August 1956, "I can think of nothing that would have a more sobering effect than a revival of talk in Washington about the possibility of a sea level canal across Nicaragua," which would force the Panamanians "to accept gracefully the many benefits they are now reaping."[22] While such a strategy might backfire, the official responded, the United States could quietly invoke the 1914 canal treaty with Nicaragua by sending in a surveying team: "If word of this action is picked up in Nicaragua by the Panamanians through their own intelligence channels the desired effect may be obtained."[23]

Eisenhower authorized his secretary of state to initiate talks with Nicaragua, "pointing out that already ships were being built too big to go through the Panama Canal and that a sea level canal through Nicaragua would be much more practical."[24] Such views indicated the technological and political appeal of a streamlined seaway, but also ignorance of isthmian geography. Engineers had long since deemed locks necessary for the route through the high-level Lake Nicaragua. While not impossible, a sea-level channel would require the draining of the lake, one of Nicaragua's most important natural resources.[25]

A follow-up assessment in November 1957 revealed little enthusiasm among the Joint Chiefs of Staff and the State Department for assuming the enormous cost of either a second canal or converting the existing one.[26] Even so, during the late 1950s, the PCC, the House of Representatives Committee on Merchant Marine and Fisheries (which oversaw appropriations for the canal), and other interested parties continued to investigate the question of canal modernization. Despite its decline as a strategic asset, the waterway remained a useful conduit. Equally if not more important, the Canal Zone's military complex had become a major instrument of hemispheric security by housing telecommunications and other means of monitoring Central America.[27]

"A storm is building up in Panama," warned the perceptive authors of the April 1959 *Foreign Affairs* article.[28] On May 1, university students surprised U.S. officials by planting several dozen Panamanian flags in civilian spaces of the Canal Zone. "Operation Sovereignty," as the students called it, broadcast their demand to end U.S. control over the Zone. White Zonians dismissed the demonstration as a joke, but later in the month, riots broke out, marking the second year in a row of lethal anti-U.S. protests in Panama.[29]

As gestures of goodwill, the outgoing Eisenhower administration issued an economic aid package and an executive order that the Panamanian flag be raised beside the Stars and Stripes at a conspicuous site in the Zone bordering Panama City. But again, change proceeded slowly due to court challenges and congressional resistance to reforms that might dilute U.S. dominance over the Canal Zone.[30] Representative Flood spoke for many of his colleagues by denouncing the 1936 and 1955 modifications of the 1903 treaty as "a piecemeal liquidation of our sovereign rights, power, and authority on the isthmus." Concessions served only to encourage more acts of "irresponsible political extortion" by Panamanian radicals and Communist agitators. The recent victory of Fidel Castro's Soviet-supported revolution in Cuba compounded conservative U.S. fears of "a rising Red tide" converging against the Hay–Bunau-Varilla Treaty, the Monroe Doctrine, and the United States of America itself.[31]

It was in this context of increasing tensions between the United States and Panama, between U.S. advocates and opponents of making concessions to Panama, and between the U.S. and Soviet systems—which competed for influence and allies in the developing world—that the "Panama Canal II" proposal reemerged with a technological twist.[32]

Plowsharing Alaska and Panama

The Suez Crisis had manifested the threat to Europe's empires posed by the postwar decolonization movements, and the temporary closure of the canal divulged the dependence of European and Israeli consumers on petroleum, tea, and other strategic commodities transported via the Red Sea–Mediterranean shortcut.[33] For an elite group of scientists at the Lawrence Radiation Laboratory in Livermore, California, the Suez closure sparked a different realization. What if atomic energy could be harnessed to cut an alternative route through "friendly territory"?[34]

The idea of using nuclear power for nonmilitary purposes, such as generating electricity and treating cancer, had already taken root via Eisenhower's "Atoms for Peace" program.[35] Applying the atom's explosive power to megaproject planning became the central goal of Project Plowshare, which the Atomic Energy Commission, the federal agency responsible for developing and promoting nuclear power, initiated in July 1957. PNEs would provide not only an economic boon to the construction industry and a diplomatic option for decision makers facing transportation crises like the Suez closure but also job security to nuclear scientists and engineers if the superpowers succeeded in negotiating treaties mandating disarmament, an important geopolitical development of the late 1950s. Indeed, Project Plowshare provided a rationale for the Livermore facility to continue operating while the United States observed a voluntary moratorium on nuclear testing from November 1958 to September 1961.[36]

Project Plowshare's most visible advocate was Livermore's cofounder, Hungarian-American physicist Edward Teller. Teller had achieved fame for his foundational scientific and controversial political roles in the development of nuclear weapons during and after World War II. He rejected the qualms of fellow physicists regarding the development and stockpiling of nuclear weapons and considered the dangers of radioactive fallout from nuclear weapons testing to have been "greatly exaggerated."[37]

Teller's vision for nuclear civil engineering included many applications, including fracturing underground rock formations to harness natural gas, heating tar sands to recover oil, and creating isotopes for medical purposes. The most dramatic uses of PNEs pertained to earthmoving endeavors. Canals, harbors, reservoirs, mountainous roads, and other such products of "geographical engineering" could be quarried at a fraction of the cost and risk of conventional explosives. Geographical engineering was in essence the Atomic Age version of the utilitarian conservationist "gospel of efficiency"—the philosophy of subjecting natural resources to technocratic oversight and development in order to reduce waste and provide the greatest good for the greatest number.[38]

The concept of PNE-facilitated public works was "very simple," as Teller's colleague Gerald Johnson explained to the chair of the congressional committee responsible for nuclear energy policy-making: "A ditch is constructed by detonating a sequence of buried nuclear explosives so spaced as to provide a smooth channel. The explosion is used not only to shatter the material, but also to eject it from the cut. In this way the desired excavation is accomplished in a single step."[39] Although everyone within a ten-thousand-square-mile fallout area might have to be evacuated for a year, excavating a six-hundred-foot-wide waterway by nuclear means might reduce costs by 84 percent. Even if the preliminary calculations were off by as much as 50 percent, "the savings would still be substantial."[40]

The desire to demonstrate the feasibility of large-scale nuclear excavation on U.S. soil led Teller to promote Project Chariot, an initiative to use five buried nuclear bombs equaling five hundred thousand tons of TNT to create an "instant" harbor in northwestern Alaska.[41] As discussed by several scholars, Chariot involved a great deal of hubris and miscalculation on the part of the AEC from the project's start in 1958 until its demise four years later.[42]

Teller and Johnson, Plowshare's first director, underestimated resistance to Chariot by Alaskans, who distrusted the claim that "all but a very small percentage of the radioactivity will be safely contained underground."[43] To allay such concerns, the AEC offered grants to University of Alaska biologists for baseline studies of the coastal Arctic tundra's ecological and human communities. Ecological and anthropological insights would help determine the optimal time of year to detonate the explosives so as to limit exposure to radiation. Ideally, the lucrative grants would also provide authoritative endorsements for the nuclear harbor by local scientists.[44]

For John Wolfe, the founding director of the AEC's new Environmental Sciences Division, the Alaskan research program represented a priceless opportunity to conduct a predetonation biological survey. At Plowshare's second symposium in 1959, the former professor of botany emphasized the uniqueness of the Arctic ecosystem, a time when most outsiders considered it a wasteland. He also urged his colleagues to heed the advice of the conservationist Aldo Leopold to recognize "the complexity of the land organism."[45] In the 1930s and 1940s, Leopold had challenged the utilitarian approach to conservation by arguing that humans must appreciate nature as a life-giving system of interdependent ecological relationships, not merely as a set of commodities to be managed.[46] Wolfe even argued for applying Leopold's ethical framework to the sea, long before most biologists expressed concern about oceanic health.

Wolfe concluded his Plowshare symposium presentation on an unexpected historical note. The idea of preliminary biological surveys was nothing new, he argued; it underpinned the famous Lewis and Clark Expedition of the early 1800s. After acquiring a vast landscape extending westward from the Mississippi River, President Thomas Jefferson had instructed Captain Meriwether Lewis to collect data on the natural history and Indigenous peoples of the Missouri and Columbia River basins. While conceding that the expedition achieved few of Jefferson's scientific objectives, Wolfe deemed the inspiring intellectual quest applicable to the Alaska project.[47]

As the Chariot researchers quantified food webs and other aspects of coastal Arctic ecology, the Livermore physicists three thousand miles to the south worked with engineers of the PCC and a special subdivision of the Army Corps of Engineers, the Nuclear Cratering Group, to investigate the feasibility of building a nuclear seaway even farther south.

Their joint report, completed in 1960, updated the 1947 PCC study by focusing on five routes that appeared most amenable to nuclear geographical engineering: in Mexico, along the Nicaragua–Costa Rica border, in eastern Panama along two separate routes, and in northwestern Colombia. Although none of the report's economic analyses included estimates for acquiring land or securing treaty rights, the construction costs of nuclear excavation appeared very favorable. The cheapest conventional option, of converting the existing canal to a six-hundred-foot-wide seaway, was $2.3 billion—three times as much as the least expensive nuclear sea-level canal, a thousand-foot-wide channel to be blasted out along the Sasardi-Morti route in the Panamanian portion of the Darién, 110 miles east of the existing waterway. While requiring precise meteorological conditions and the temporary evacuation of thousands of residents, the $770 million project appeared feasible and safe. Accordingly, the report urged the AEC to continue developing cleaner nuclear explosives with an eye toward using them for a seaway outside the Canal Zone.[48]

Around the same time, the House Committee on Merchant Marine and Fisheries released its own update on the canal modernization issue. The 1960 report, authored by a group of engineers that included General Leslie Groves, the former director of the Manhattan Project, investigated many aspects of the Panama Canal's long-term viability. The engineering consultants called for more research into isthmian sea-level canal routes and new conventional and nuclear methods of canal construction. However, they warned, "As of now, the only hope for an economically justifiable sea-level canal appears to be by excavation through as yet unproven nuclear means." Until safe, cheap PNEs could be developed, they recommended making interim improvements to the existing, aging canal as soon as possible.[49]

Only the last recommendation satisfied Representative Flood, who considered the seaway idea an artifact of postwar hysteria over atomic attacks. Now that the even more powerful hydrogen bomb had exposed "the underlying fallacies in the 'security' thesis for planning navigational projects," as he asserted in numerous speeches, he argued that policy makers should resume the World War II–era expansion project.[50]

Despite the opposition of Flood and other congressmen, in January 1961 the National Security Council issued a policy guidance report identifying the sea-level canal as an important project. To complete it by 1980, its location and means of construction "must be made soon." Comprehensive feasibility studies should thus proceed, including of the "physical, biological and psychological effects of nuclear explosives under conditions to be encountered at the canal site."[51]

For the new president, however, the seaway proposal provided a way to postpone hard decisions about U.S.-Panama relations. John F. Kennedy's administration tried to dissuade President Roberto Chiari of Panama from broaching the subject of treaty reform, but Chiari insisted the time had come for abrogating the 1903 pact. As he wrote in a September 1961 letter that soon became public, "There is no place in the mentality of man in this second half of the 20th century for the proposition that a state, no matter how strong, can exert sovereign rights over any part of the territory of another state, no matter how small or weak."[52]

Kennedy stalled in penning a comprehensive response to Chiari for seven months while another National Security Council working group reexamined the Panama Canal's current and future needs.[53] The resulting report recommended deferring any decision on a sea-level canal for five years, while conducting feasibility studies and delaying formal treaty negotiations affecting the existing waterway during that period. As for containing Panamanian pressure, the working group advised redirecting attention to the sea-level canal studies, reinterpreting the 1903 treaty "to satisfy Panamanian aspirations which are reasonable and consistent with the basic interests of the United States," and providing more economic aid to reduce dependence on canal revenues.[54]

In conveying the report to Kennedy, the State Department's second-highest official, George Ball, issued two warnings. The United States must take care not to oversell PNEs for canal construction, since such an approach could "provide a golden opportunity for Soviet propaganda throughout Latin America." Delay tactics might also backfire given the growing pressure in Panama for change and the concomitant resistance in the U.S. Congress against any dilution of U.S. sovereignty in the Canal Zone.[55] Ball's admonitions informed an ensuing National Security Action Memorandum. The confidential document directed the AEC to determine the feasibility of nuclear excavation within five years and to participate in a joint sea-level canal research program with the Army Corps of Engineers that would include prompt on-site surveys in Panama and Colombia.[56]

On the same day of the National Security Action Memorandum's internal release, April 30, 1962, Kennedy finally responded to Chiari. The timing was critical because Panama's academic year commenced in May, and Chiari's government feared that the new school year would reignite leftist student activism.[57] Kennedy invited the Panamanian president to Washington for a state visit in June, but asserted that treaty negotiations must await the completion of seaway-oriented studies over an unspecified "period of years."[58] After the visit, the two nations formed a task force to examine Panamanian grievances, but the committee kept the focus on symbolic matters, such as the postage stamps and flags used in the Canal Zone.[59]

Nineteen sixty-two did not constitute the turning point Panamanian treaty reformers had hoped for, but it did result in major changes for Project Plowshare. On the one hand, the Chariot initiative ground to a halt. The AEC had acquired the scientific data it sought, but not the concomitant political support. Rather than keeping quiet after providing the contracted information, the University of Alaska biologists worked with grassroots organizers to publicize their concerns about how the project's radioactive fallout might affect Arctic food chains. Organizers in Alaska and the continental United States generated so much negative publicity that the AEC called off Chariot after four frustrating years.[60]

It was a heavy blow for officials who viewed the harbor project as an ideal demonstration project for the nuclear seaway in Central America. And yet during the same year, the United States lifted the ban against testing nuclear weapons. Plowshare administrators wasted no time in conducting thermonuclear cratering experiments at the Nevada Test Site. The first one, Project Sedan, took place in July 1962. Sedan, a hundred-kiloton device buried 635 feet underground in alluvium, emitted a mushroom cloud that could be seen 65 miles away in Las Vegas. Its massive crater—1,280 feet in diameter and 320 feet deep—is now listed on the National Register of Historic Places. Within two hours of the shot, the AEC announced that 95 percent of the radioactivity had been contained.[61] Yet elevated levels of iodine-131 soon appeared in milk supplies in Salt Lake City, Utah. AEC officials assured worried public health officials that everything would be okay once dairy farmers shifted their cows from fallout-contaminated pasture to dry feed.[62]

Over the next year, high-profile publicity reiterated the agency's confidence in its ability to develop so-called clean explosives for nuclear excavation.[63] Meanwhile, bills to fund detailed technical site surveys along the proposed Central American routes languished in Congress throughout 1963, as members debated who should oversee the feasibility study—the agencies with a vested interest in new construction, top administration officials such as the secretary of state and secretary of defense, private citizens, or various combinations thereof.[64]

Representative Herbert Bonner, the chair of the Committee on Merchant Marine and Fisheries who had commissioned the 1960 House canal report, introduced a bill to authorize the PCC to investigate methods of improving the canal's security and capacity or of building a new channel to address future commercial and defense requirements. By contrast, Representative Flood demanded an investigative commission independent of the PCC and Army Corps of Engineers, and advocated for an adapted version of the suspended third-locks expansion of 1939–42 so as to maintain U.S. territorial rights in the Canal Zone. He blamed the failure to modernize the present canal on procrastinating officials and the few stakeholders who stood to benefit from a sea-level canal—earth-moving machinery manufacturers and military and civilian engineers hoping to gain long-term contracts.[65]

Despite his nationalist biases and blind spots, Flood made valid points about the seaway proposal's limited nongovernmental support. A key potential user, the shipping industry, sought to keep tolls as low as possible and distrusted claims that a new sea-level waterway would remain as cheap to transit as the Panama Canal, an issue that would be explored in much more detail by the succeeding administration. Furthermore, the Livermore laboratory and AEC had a strong stake in finding new projects as public support increased for nuclear arms control. In a similar vein, the Army Corps of Engineers had a well-deserved reputation as a powerful Washington lobby that leveraged congressional relationships to ensure an endless pipeline of projects.[66]

Another action-forcing event in 1963 publicized the nuclear canal proposal: the congressional hearings regarding an international agreement to ban nuclear weapons tests in the atmosphere, in outer space, and under water. By that time, U.S. and Soviet physicists had long since developed bombs that dwarfed the explosive power of the ones deployed against Japan. In 1954, the AEC tested a thermonuclear bomb at Bikini Atoll with an astounding yield of fifteen megatons of TNT. The experiment unleashed radioactive fallout over a much wider swath of the Pacific Ocean than expected and sickened the members of a Japanese fishing boat, generating worldwide outrage. No matter how remote test sites appeared to be, the dozens of experimental detonations conducted each year by the three nuclear powers—the United States, United Kingdom, and Soviet Union—released radioactive isotopes that made their way into human bodies. Campaigns to expose the public health effects of radioactive fallout helped drive the diplomatic efforts to decelerate the nuclear arms race during the late 1950s and early 1960s.[67]

While allowing underground tests to continue, the proposed test ban treaty posed an existential threat to PNE projects in smaller nations by banning cross-border releases of radionuclides. Because buried nuclear charges might vent enough radiation to cross hundreds of miles into adjacent nations, how could an interoceanic canal be built in any of the countries spanning the Central American isthmus?

When asked at one of the hearings whether the United States had "any immediate plans to begin exploding atomic energy to build canals or to build harbors or to blow up mountains," Atomic Energy commissioner Glenn Seaborg admitted, "We are not ready." Nevertheless, he asserted, excavation technology *experiments* could proceed under the proposed treaty, and future nuclear construction projects could take place as long as the parties to the treaty agreed to amend it.[68]

In October 1963, the U.S. Senate ratified the Limited Nuclear Test Ban Treaty, a milestone for international arms control, Cold War diplomacy, and environmental health protection. But it upset advocates of geoengineering because the final document, unlike earlier versions, included no exceptions for PNEs, not even on an experimental basis.[69]

Cautious optimism prevailed nevertheless that agreements could be worked out with the Soviets, who were pursuing their own PNE projects, to allow cratering shots to continue.[70] Moreover, as Plowshare director Gerald Johnson assured the chair of the congressional Joint Committee on Atomic Energy, in the three years since the completion of the 1960 report, researchers had reduced both the cost estimates of PNEs and projected levels of radioactivity in fallout. The technical, economic, and military rationales for the sea-level canal appeared ripe for reanalysis.[71]

Conclusion

Midcentury technological and economic innovations reduced U.S. reliance on the Panama Canal as a commercial and military conduit, but by then the Panama Canal Zone had become a critical locale for Latin American–focused surveillance, security, and defense infrastructure. Accordingly, except for a few modifications, postwar U.S. officials resisted making fundamental changes to the Hay–Bunau-Varilla Treaty.

Yet maintaining hemispheric security interests while managing the more obvious issues of canal ownership and operation was becoming more and more difficult. The unequal living standards and employment privileges enjoyed by U.S. civilian canal employees and their families, and many other instances of injustice, bred bitterness and violent demands to revise the 1903 U.S.-Panama treaty.

The proposal for a sea-level canal, especially one constructed with the new technology of PNEs, appeared to address the technopolitical outdatedness of the Panama Canal and its zone. For Presidents Eisenhower and Kennedy, the seaway idea, in concert with symbolic concessions, offered a background kind of diplomatic leverage to placate Panamanian dissent indefinitely. Because it was such a technically complex undertaking, the proposal bought time for Plowshare scientists and engineers to develop safer explosives and build greater public trust, and for U.S. officials to improve diplomatic relations with Panama without provoking the many Americans who considered the Canal Zone their own.

In hindsight, each of these endeavors was as riddled with unsustainable, contradictory premises as the Canal Zone system itself. Nevertheless, they reinforced the exciting conditions of possibility embodied by the nuclear seaway of the future, a project designed to imbue the Panamanian landscape with multiple forms of modernity.

CHAPTER 3

Mobilizing for Panama Canal II

NINETEEN SIXTY FOUR, the fiftieth anniversary of the completion of the Panama Canal, was anything but golden. What was to have been a year of celebration opened instead with four days of anti-U.S. violence across Panama, worldwide coverage of which featured U.S. soldiers firing on citizens of their host country. Panama's president, fed up with the delays over treaty reform and eyeing an uphill election campaign for his conservative party, broke diplomatic ties and refused to resume them without an agreement to abrogate the 1903 accord that granted the northern colossus perpetual control over the waterway and its surrounding zone.

While the new U.S. president was not about to bullied, he knew the status quo could not endure. Lyndon B. Johnson had assumed office following the tragic assassination of Kennedy on November 22, 1963. That morning, Plowshare and State Department officials had met to discuss how the nuclear sea-level canal proposal could remain viable in the context of the Limited Nuclear Test Ban Treaty, signed by Kennedy a month earlier.[1] The January Flag Riots, as they became known, led Johnson to resume secret discussions over the political benefits of building a nuclear canal on the cheap in either Panama or another nation. He sought to defuse Panamanian demands for treaty reform until after the November 1964 U.S. presidential election. Johnson had to tread a fine line between managing dissent among frustrated Panamanians and the demands of anticommunist U.S. interests to take a hard line against the small yet strategic nation.

The sea-level canal provided the ideal venue for doing so, and Johnson masterfully used the proposal to contain both foreign and domestic opposition during the pivotal year following the Flag Riots. However, it is not necessary to depict the nuclear seaway as a "peculiar futuristic fantasy" whose "fictitiousness" allowed him and his opponents to make of it what they wanted.[2] Such language overlooks the important *work* that went into the feasibility studies—work of both a political and scientific nature—by people who took the existing canal's obsolescence, and the technoscientific innovation of peaceful nuclear explosives, very seriously. Part II will address the feasibility investigations in detail, and this final chapter of part I addresses the hidden political groundwork for the law creating the multimillion-dollar Atlantic-Pacific Interoceanic Canal Study Commission and for the historic presidential announcement in December 1964 regarding the renegotiation of the 1903 Hay–Bunau-Varilla Treaty.

Permeating these two developments was an intriguing degree of what could be called technological anti-imperialism. Weapons and other tools have enabled their possessors to control other territories throughout history, and during and following the Spanish-Cuban-American War, the United States used its technological superiority to build a tropical empire across the Caribbean Sea and Pacific Ocean—often under the guise of civilizing island peoples.[3] Yet rarely have powerful nations used technology as a vehicle or pretext for decolonization, which makes the Johnson administration's deployment of the sea-level canal proposal all the more remarkable.

Johnson harnessed the disparate economic, military, and political forces favoring a modern new waterway to nudge forward the cause of Panamanian sovereignty over the Panama Canal Zone. A streamlined ship channel of 1,000 feet wide and 250 feet deep would not only accommodate the world's largest vessels and decrease transit times, it would also preclude the need for an adjacent enclave populated in perpetuity by foreign canal operators and defenders. Johnson's two predecessors, Eisenhower and Kennedy, had also viewed the sea-level canal proposal as a means of dismantling the problematic Canal Zone, but Johnson worked to get the feasibility studies going in a much timelier manner.

That is not to say that Johnson did not employ delay tactics, nor that his vision of a colony-free canal lacked bias. The idea that a simplified seaway would enable its owners to dispense with highly skilled personnel recapitulated the patronizing presumption that few if any Panamanians could be trusted or trained to operate complex infrastructure, one of the many points of contention between the two nations. Moreover, the administration sought to reduce Panamanian leverage over the treaty negotiations by calling for the investigation of sea-level canal routes in other countries besides Panama.

Despite its important anticolonialist components, the seaway proposal undermined principles of equity and sovereignty in other ways. Johnson did not seek to withdraw all U.S. personnel from Panama, due to the region's increasing postwar role in hemispheric surveillance, military staging, and jungle guerilla-warfare training—developments exposed by the Flag Riots.[4] New diplomatic arrangements would be needed to bring Panama's U.S. bases in line with the leasehold agreements established for hundreds of other overseas facilities run by the Pentagon during the Cold War. Accordingly, the Johnson treaty negotiators sought separate pacts governing the proposed new seaway and the existing military bases.

Finally, the progressive goal of recognizing Panama's full sovereignty over its lands via a new sea-level canal treaty must be considered in the context of contemporaneous foreign policy initiatives that exuded technological arrogance. Johnson officials funded a secret experimental program to control the weather in India and Pakistan for food production and drought mitigation.[5] Much more famously and disastrously, the administration escalated the anticommunist war in Vietnam despite intelligence assessments that superior firepower could not overcome the enemy's resolve.[6] Johnson also expanded his predecessor's program of chemical herbicidal warfare.[7]

These cases demonstrate how Johnson embraced science and technology as diplomatic instruments to control and reengineer the physical environment and less technologically advanced nations. It is no mere coincidence that the major architects of the Vietnam War—Secretary of Defense Robert McNamara, National Security Advisor McGeorge Bundy, and his successor Walt Rostow—also championed the nuclear Central American canal. The Johnson administration's advocacy of the nuclear seaway contributes to scholarship on the foreign policy roles played by science and technology during the Cold War, especially with respect to large-scale environmental transformations.[8]

A Tale of Two Treaties

Throughout 1963, while the U.S. Congress debated the terms of the Limited Nuclear Test Ban Treaty, anti-Yankee tensions intensified in Panama. Facing anger from young Panamanians, the oligarchic Chiari renewed his public demands for a new treaty, but the U.S. government would commit only to constructing new flagpoles in the Zone to fly the Panamanian standard alongside the U.S. one. That and other symbolic concessions infuriated conservative members of Congress; as their spokesman Representative Dan Flood contended, Panamanian flagpoles in the Canal Zone signified yet another stage of "the long-range Soviet program for conquest of the Caribbean."[9] He was referring to the Cuban Missile Crisis, the thirteen-day standoff in October 1962 that almost led to a nuclear war between the United States and the Soviet Union. The Soviets had deployed ballistic missiles to Cuba following the failed 1961 U.S. plan to overthrow Fidel Castro, the leader of the communist Cuban Revolution. Although the Soviet and U.S. leaders negotiated a resolution, for almost two decades afterward the crisis fueled Flood's campaign to maintain U.S. control of the Panama Canal Zone.

In response to Flood's resolution to bar the flying of any flag in the Zone other than the Stars and Stripes, the syndicated columnists Rowland Evans and Robert Novak penned an editorial titled "Ugly Americans." If anti-U.S. sentiment ever reached the point of threatening the canal, the authors argued, the real culprit would be not Communist agitators but rather the U.S. policy makers who perpetuated "big city imperialism" and paid Panama a "chickenfeed" annuity from the annual canal tolls (less than $2 million out of $60 million in revenue). They also targeted imperious Zonians who acted as "badwill ambassadors" and "one of Washington's most effective lobbies by playing on justifiable congressional fears about this vital waterway's security."[10] However, as events would soon make clear, most Americans took great pride in U.S. ownership of the canal and zone.[11]

In early 1964, the governor of the Canal Zone, Robert Fleming, ordered that the number of sites flying U.S. flags be limited, and that wherever the U.S. flag was flown, the Panamanian standard must accompany it. Because Balboa High School, which was located in a part of the Zone near Panama City, had only one flagpole, its leaders took down the banner to avoid violating the order. Encouraged by their parents, several Zonian teenagers raised a makeshift replacement on the night of January 7–8, 1964. Word spread, and outraged Panamanian students marched into the Zone on the ninth to raise their own flag, a bloodstained relic of a 1947 riot. A fight broke out, during which the flag was ripped. Radio broadcasts of the incident sparked uprisings in Panama City and Colón. In the violence that consumed the country on the ninth, tenth, eleventh, and twelfth, when the Guardia Nacional finally resumed control, twenty-one Panamanians and four U.S. soldiers died. President Lyndon B. Johnson was caught off guard by the violence and Chiari's suspension of diplomatic relations, the first foreign relations emergency of his new administration.[12]

The Flag Riots cast an embarrassing light on U.S. imperialism in Panama. As an editorial in the glossy *Life* magazine stated, "As owner and operator of the Canal Company, the U.S. government has blindly allowed the Canal Zone to turn into a pretty fair imitation of a colony, complete with a colonial mentality. In the Zone, discrimination against Panamanians has existed since the beginning, backed up by wage differentials, special privileges for Americans and all the paraphernalia of extra-territoriality."[13] Many such privileges had nothing to do with the canal, including access to U.S.-operated businesses that kept prices low for Zonians and undercut Panamanian entrepreneurship.[14] More than 36,000 Americans lived in the Zone in 1964, including 9,750 active duty military personnel, 11,800 military family members, 3,905 civilian employees of the armed forces and their families, and 10,700 Panama Canal Company and Zone government employees and their families. The PCC also employed 5,000 Panamanians, and the U.S. military employed about 10,000 Panamanian citizens.[15]

Similar critiques appeared in domestic and international media, and Panama's National Bar Association asked the International Commission of Jurists, a non-governmental organization based in Geneva, Switzerland, to investigate whether U.S. actions in Panama from January 9 to January 12 violated the United Nations Universal Declaration of Human Rights.[16] However, while Johnson officials worked to contain the damage, 56 percent of Americans demanded that their government make no concessions to Panama.[17]

The day after the riots ended, Johnson met with several high-level officials including Secretary of State Dean Rusk, Secretary of Defense Robert McNamara, and Assistant Secretary of State for Inter-American Affairs Thomas C. Mann to discuss solutions to the diplomatic impasse. Mann, a controversial diplomat whom Johnson had appointed the month before, argued for using the threat of building a sea-level canal in Colombia or Nicaragua as a means of leverage with Panama so as to develop a new treaty acceptable to the U.S. Senate. Johnson and officials from the State and Defense Departments and Atomic Energy Commission agreed, but the Limited Nuclear Test Ban Treaty of 1963 posed a daunting obstacle. By prohibiting cross-border releases of radioactive debris, the treaty diluted the threat to Panama of building a nuclear canal in another country.[18]

Another strong Johnson administration advocate of using the nuclear canal option to manage both domestic opposition and Panamanian demands regarding the renegotiation of the 1903 treaty was Deputy Secretary of Defense Cyrus Vance, who in his former role as secretary of the army had held the distinction of being the Panama Canal's sole shareholder.[19] Vance warned the president on February 10 that both hard-right and moderate Republicans might seize the opportunity to their advantage in the lead up to the November election: "[Barry] Goldwater, [Nelson] Rockefeller and others may well raise the cry that in negotiating the Test Ban Treaty, the Democratic Administration precluded the sea level canal solution to the Panamanian problem, or at least precluded the more desirable method of construction." To contain the damage, he suggested advance surveying work could begin while negotiators obtained a test ban treaty exemption for peaceful nuclear excavation in smaller countries. Vance urged Johnson to support a bill that had been pending in the House for over a year to provide $17 million for the PCC and Army Corps of Engineers to survey two remote Darién routes, one in Panama and the other in Colombia, to determine the optimal site and method of construction. Otherwise, if further delays ensued, "the political opposition may well push for the site surveys themselves. They then enjoy the best of all worlds. They can criticize our handling of the Test Ban Treaty and claim credit for initiating the only constructive program to solve the Panamanian problem."[20] Although Vance's concerns about getting scooped did not materialize, the ensuing 1964 Republican platform did call for studying the feasibility of a sea-level isthmian canal with nuclear methods.[21]

The Flag Riots stimulated bipartisan congressional support for nuclear canal excavation, and thus the AEC pressed the White House for permission to resume relevant work at the Nevada Test Site. It had been nineteen months since the first and last excavation experiment, Project Sedan, although other Plowshare detonations oriented toward device development had taken place there in the meantime. In contrast to the hundred-kiloton Sedan device that left an enormous crater on the desert floor, the next proposed excavation test, Project Sulky, involved a blast of only ninety-two tons. But concerns that the explosion might still vent enough radionuclides to be detected in Canada or Mexico sparked an internal debate about the risks of violating the test ban treaty. The administration decided in February 1964 to postpone the project until the following winter. The intervening months would provide more time to continue developing less-contaminating explosives and thereby improve the prospects of nondetection, time to ensure the most advantageous wind and weather conditions, and time to address the potential for amending the Limited Nuclear Test Ban Treaty. By then, the grazing season and presidential election would have also passed, thereby reducing the test's literal and figurative fallout.[22]

The long-delayed congressional bills for sea-level canal feasibility studies got a new round of hearings in March 1964. Cabinet officials framed the nuclear seaway as the solution to long-standing problems encompassing the technological, economic, and political realms. Vance reiterated the postwar history of calls for canal modernization, and explained "the current difficulties" between the two nations as a function of the lock waterway's complexity and concomitant need of a large U.S. labor force. As Mann elaborated, "I understand that the present canal . . . has some 14,000 employees, and a sea level canal would only require some 600, and presumably only a fraction of that number would need to be there [twenty-four hours a day, seven days a week]." Further testimony established that the six hundred people need not be U.S. nationals and that fewer military personnel would be needed to defend a canal lacking "the complex lock situation that you have now."[23] Subsequent hearings considered the question of who to appoint to the commission responsible for determining the feasibility of nuclear excavation.

The tense suspension of diplomatic relations between Panama and the United States ended on April 3 with a joint declaration to work toward "the prompt elimination of the causes of conflict between the two countries."[24] The same day, Johnson appointed lawyer Robert B. Anderson, a Texas Republican who had served as secretary of the navy and secretary of the treasury under President Eisenhower, to direct the team in charge of the talks. Anderson had come highly recommended by Mann to serve as the "tough guy" in charge of the "hard negotiating" with the Panamanians (though Mann later called the recommendation one of the "great mistakes" of his life).[25] Johnson also liked to appoint Republicans to controversial posts to deflect heat from himself.[26] The Panamanian officials with whom Johnson administration officials had conferred did not want a career diplomat heading the team, and Anderson attributed his acceptability to the goodwill he had established among Panama's ruling elite by leading the 1959 effort to organize the Inter-American Development Bank. Looking back in the late 1970s, Anderson noted Johnson's dedication to restoring diplomatic relations with Panama but also his unrealistic assumption that the treaty negotiations could be completed within two to three months—a far cry from the two-and-a-half years over which they stretched. Anderson also recalled that in their initial discussion, Johnson expressed concern about how the Panama Canal related to one of his signature initiatives, the space program, for which some of the enormous rocket parts had to be transported via barge from California to the Florida launch site.[27]

Anderson agreed to serve as the special representative for U.S.-Panama relations in exchange for access to the president and permission to remain based in Manhattan, and Johnson established two new entities to formulate and execute U.S. policy in Panama, one in Washington and the other in Panama. The Washington-based Panama Review Group met on April 7 to discuss Anderson's negotiating strategy and unilateral actions that could be taken without congressional approval "in order to blunt interim Panamanian pressures, and hold the line until early 1965 when both Panamanian and United States elections would be behind us." Stephen Ailes, who had recently replaced Vance as the secretary of the army, emphasized the need to "dispel our colonialistic image." He called for securing an agreement for new canal site surveys in Panama, countering Panamanian concerns that a second waterway would ruin their economy, and otherwise deflecting attention from the " 'perpetuity' problem and other highly charged issues" raised by the existing Canal Zone establishment: "We have to make a really penetrating study. Panama's demands on basic issues should be countered by tying them into the sea level canal rights we will want."[28]

The other attendees agreed that framing the upcoming treaty negotiations in the context of a new sea-level canal would help distract from the "present situation," while recognizing the need to negotiate separate military base rights.[29] The Philippines offered a precedent for such an arrangement, since the U.S. retained naval and air facilities there after recognizing the country's independence in 1946. Moreover, the U.S. had negotiated postwar military base agreements with nations around the world, most of which involved leases rather than the permanent status granted by the Hay–Bunau-Varilla Treaty.[30]

In June 1964, the Swiss-based International Commission of Jurists reported its findings regarding the question of whether the U.S. response to the Flag Riots constituted human rights abuses. The jurists concluded that Canal Zone police and U.S. Army personnel used disturbingly excessive force at some points, but that the use of force per se was justified to quell the riots. While they did not violate human rights, Canal Zone authorities and police "could have handled the situation [at Balboa High School] with greater foresight," and henceforth the United States should "take effective steps to make possible a reorientation and change in the outlook and thinking of the people living in the Canal Zone."[31] U.S. officials had already agreed among themselves that "visible evidence of progress" in U.S.-Panama relations was needed prior to the Panamanian Independence Day holiday of November 3, "or else there will be trouble."[32] Others had amplified the long-standing progressive argument that the United States could afford to relinquish many of its "peripheral privileges" in the Zone without harming its maritime and security operations.[33]

One antiquated perk that attracted attention was the 25 percent tropical hardship differential paid to U.S. citizen employees in the Canal Zone. The salary boost had been deemed necessary to attract northern whites in the early 1900s, when malaria and yellow fever posed dangerous threats to foreigners lacking immunity. But sanitary engineering, chemical pesticides, air conditioning, and other technological advances had long since reduced the hazards of tropical living. The large allowance had the unintended consequence of creating what Ailes called "a second and third generation of U.S. workers in the Zone, who quite naturally resist any move designed to increase Panamanian employment in better paying jobs." While privately criticizing U.S. canal employees "who regard jobs in the Zone as matters of right," Ailes stated that his proposed reduction in the hardship differential to 15 percent of base pay should be framed as a cost-saving matter "completely divorced from our negotiations with Panama."[34]

By August 1964, the Canal Zone government had implemented several measures designed to improve relations with Panama. The changes included installing dual flagpoles at all Zone schools, increasing wages for Panamanian employees, hiring Panamanians for the Canal Zone police force, reducing the number of jobs reserved for U.S. citizens, desegregating swimming pools and government housing, nominating a Panamanian to serve on the PCC board of directors, providing scholarships for Panamanians to attend the Canal Zone College, and proposing the hardship pay decrease. In addition, the fiftieth anniversary of the opening of the waterway transpired with "quiet and restrained ceremonies which were not offensive to Panama," as the secretary of defense assured the president.[35]

On the other hand, the Zone leadership overlooked other requested changes, such as having ships transiting the canal fly the Panamanian flag (in addition to the U.S. one and the flag of registry), making Spanish an official language, and using the host country's postage stamps. These and other points of contention generated anxious discussions among Johnson officials months later, as the one-year anniversary of the Flag Riots approached and threatened to unleash anti-American demonstrations "of even more serious proportions."[36]

Experiment No. 1

U.S. officials hoped the concessions would lessen Panamanian ire while they negotiated long-term plans for a new, zoneless seaway.[37] In the August 1964 memorandum assuring the president of the beneficial changes in the Zone and the subdued golden-anniversary commemorations, Secretary of Defense McNamara concluded, "The best prospect for a major improvement in U.S.-Panamanian relations is that offered by the sea level canal project." Echoing the rationales made by Vance, Mann, Ailes, and others, he explained that an agreement for a new seaway in Panama "would put to rest many of the emotional issues which now plague our relations. It would also clear the air of many of the uncertainties with respect to United States policy which are the source of most of the unrest among the U.S. citizens in the Zone."[38] McNamara's coded language spoke to the discomfort felt by senior U.S. officials about the disruptive behavior of both extremist Panamanians and Zonians. A new treaty for a simplified waterway requiring far fewer U.S. employees—who would be required to leave by a fixed date, even if it were far in the future—held the key for improving relations with Panama over both the short and long terms. Toward that end, a new feasibility study authorization bill that had emerged from the March hearings (S. 2701) had passed the Senate and was scheduled for an upcoming House vote.[39]

The hearings had relied on old data regarding the feasibility of nuclear excavation, as did all the news coverage following the Flag Riots of the possibility of an "atom-dug" canal.[40] But the administration's efforts to buy time prior to the November presidential election had another payoff in the form of a new technical assessment of the nuclear-excavated canal, the first since 1960. The joint report by the PCC, the Nuclear Cratering Group of the Army Corps of Engineers, and AEC discussed the latest experiments with chemical and thermonuclear explosives and concluded that "major progress has been made in development of 'cleaner' nuclear explosives applicable to excavation."[41]

The report, titled *Isthmian Canal Studies—1964*, focused on the two routes deemed most promising for PNEs. Route 17, the Sasardi-Morti passage through eastern Panama's Darién region, called for detonating 294 nuclear explosive devices with an aggregate yield of 166.4 megatons, to be fired in fourteen separate detonations. While the average yield per detonation would be approximately 10 megatons, cutting through the 1,100-foot-high Continental Divide would require an explosion of some 35 megatons. By comparison, the most powerful nuclear device ever tested by the United States, the 1954 Castle Bravo blast that released radioactive fallout over a hundred-mile swath of the Pacific Ocean, yielded 15 megatons.

The other proposed nuclear path between the seas crossed Colombia's portion of the Darién isthmus via the Atrato and Truando Rivers. Designated Route 25, it closely paralleled the route that had intrigued the elderly Humboldt in the 1850s. Plowsharing the mountainous terrain would require an awe-inspiring degree of explosive power: 262 nuclear devices with an aggregate yield of 270.9 megatons, to be fired in twenty-one separate detonations.

Each of the proposed projects would require evacuating thirty thousand people due to the "main hazards from radioactivity, air blast, ground shock, throwout, and dust" and resettling them "in adjacent frontier areas." As for the cost, emplacement drilling, explosive charges, and other direct expenses of nuclear excavation would require approximately $307 million for the Panama route and $315 million for the Colombia one. To improve the economic and safety estimates, the army engineers and Plowshare scientists called for the accumulation and analysis of much more data pertaining to the topography, geology, meteorology, hydrology, coastal hydrography, seismology, ecology, demography, and economic resources of the two regions.[42]

The congressional bill authorizing the group that would be responsible for overseeing the formidable research program, the Atlantic-Pacific Interoceanic Canal Study Commission, passed both houses in September 1964. The chairman of the hearings, Senator Warren G. Magnuson, and others avowed that the legislation was "not a product of the Panama crisis." Rather, the Flag Riots had expedited long-standing calls for action regarding the obsolescing canal.[43] Although defenders of the 1903 treaty argued that there was no such thing as an invulnerable waterway, Magnuson and other Plowshare proponents insisted that "a canal built in such a manner would be almost defense proof, because any bomb landing on such a canal might make it an even better one if the bomb should blow enough dirt out."[44]

Magnuson, the chair of the powerful Senate commerce committee, was enthusiastic about PNEs (fig. 3.1). His home state of Washington was the site of the Hanford facility that produced the plutonium for the nation's nuclear weapons. At the March 1964 hearing, he had asked Seaborg the leading question, "So the urgency of this sea level canal could act almost as a laboratory, experiment No. 1, for opening a whole new vista for the use of nuclear power?"[45] Magnuson also emphasized the Soviet goal of geoengineering the Bering Strait to improve navigation and warm up Siberia: "This isn't too farfetched at all. . . . They have been talking about this for a long time, hoping that new nuclear technology can be put to some good uses. And Russian engineers have talked on many occasions about building harbors along the northern route to the Siberian coast by the use of nuclear explosives."[46]

In their quest to authorize a commission to investigate the nuclear seaway, Magnuson and like-minded members of both houses of Congress outmaneuvered their colleagues who opposed any changes to the 1903 treaty.[47] Representative Flood did succeed in ensuring that neither cabinet members nor civil servants would oversee the feasibility studies. However, the final bill authorized the president to appoint five private citizens without congressional consent.[48] Public Law 88-609 passed on September 22, just over a month before the presidential election, and eleven days before the end of the 1964 session, which had featured the historic debates over the Civil Rights Act and the Gulf of Tonkin Resolution that deepened U.S. military engagement in Vietnam. Getting the nuclear seaway commission through required intensive lobbying on the part of cabinet officials; as a staff member of the National Security Council later explained to National Security Advisor McGeorge Bundy, "We put considerable heat on the Congress to approve the legislation on an interoceanic canal commission during the waning hours of the last session."[49]

Johnson won the election on November 3 in a landslide, a victory that provided breathing space and political capital regarding domestic opposition. That day also happened to be the Panamanian holiday commemorating the 1903

FIGURE 3.1. Lieutenant General W. K. Wilson Jr., Lieutenant Colonel Robert W. McBride, and Senator Warren G. Magnuson discussing the use of nuclear charges to excavate a second Central American canal, March 12, 1964. University of Washington Libraries, Special Collections, SOC6742.

revolution against Colombia, but it did not unleash the new wave of anti-Yankee violence U.S. officials had feared earlier in the year. Chiari's oligarchic Liberal Party, now headed by his cousin Marco Robles, had won the May election, one in which U.S. officials turned a blind eye to evidence of fraud. In the weeks prior to the U.S. presidential election, Robles reversed Chiari's course of stoking anti--U.S. sentiment.[50]

Yet that was no guarantee that the upcoming one-year anniversary of the Flag Riots would pass without another international incident. Accordingly, Johnson's Panama-based policy review committee met days later to develop guidelines for the treaty renegotiations. The group deemed the top priority "a broad agreement for the possible construction and operation of a sea-level canal in Panama, a canal whose technical simplicity would permit us to forgo the rights of perpetuity and jurisdiction which we have hitherto enjoyed on the Isthmus and which have served to cloud the fundamental friendship between our two nations." The negotiators' next goals were to secure agreements for military base rights and interim changes to the existing canal operations that would provide more substantive benefits to Panama as well as "symbolic shows of sovereignty."[51]

The impending anniversary of the January 9 violence weighed on Johnson's mind when he called Mann later in November regarding the progress that had been made since the April resumption of diplomatic ties. As the phone transcriber noted, "Mr. Mann told the President that the thing that is going to help us the most is to get out in front with a lot of publicity on this new canal." By focusing on the benefits of a more efficient seaway, "we can get this whole thing in perspective, we can tell everyone here and in Panama and the whole world that the present canal is limited and that we are going to build a new one and therefore we are dealing with a wasting asset."[52] Mann assured the president he was almost done drafting a new sea-level canal treaty for Ambassador Anderson to present to the Panamanian negotiators, which helps explain why Johnson thought the matter could be wrapped up in a few months.

Indeed by December 1, Mann, working with State and Defense Department personnel, had developed the broad framework for three new treaties. The sea-level canal accord would give the U.S. the right to conduct the needed surveys and to construct the new waterway, but not to operate and maintain it; that would be the job of an international commission, and sovereignty over the seaway would be held by the country (or countries) through which it passed. The second treaty addressed U.S. military base rights in Panama, and the third one outlined the interim governance of the existing canal, over which the U.S. would relinquish control on a fixed future date: two years after the opening of the sea-level channel.[53]

Abrogating the Hay–Bunau-Varilla Treaty in such a way, explained Bundy, would secure U.S. canal rights for the time being while undercutting anti-Americanism on the isthmus. In his words, it would "remove those emotional issues (sovereignty, etc.) which provide grist for agitators in Panama."[54] Bundy's patronizing language mirrored that of other U.S. officials, even those sympathetic to the Panamanian cause, who deemed Panamanian nationalists as "emotional."[55] Of course, many Americans were equally passionate about retaining U.S. control over the Canal Zone.

Bundy urged the president to make a strong statement later in December regarding the sea-level canal treaty negotiations. If worded just right, a historic policy change in U.S.-Panama relations "would permit us to seize the initiative and dampen current efforts by anti-American elements in Panama to stage large anti-American demonstrations on January 9." To hammer out the language, the cabinet would meet the next day at the White House. Secretary of State Rusk made one significant caveat—that the possibility of nuclear excavation not be mentioned.[56]

Following further deliberations, President Johnson made his surprise announcement to the world on December 18, 1964. That was how journalists described it, but it was no surprise to former presidents Harry Truman and Dwight Eisenhower and other VIPs whose approval Johnson had secured in advance.[57] As the *New York Times* and many other newspapers reported on their front pages the next day, the United States would "plan in earnest" to replace the aging lock-based Panama Canal with a streamlined sea-level waterway and to renegotiate the 1903 treaty. Johnson outlined the technological, economic, military, and political reasons for changing course, arguing, "Such a canal will be more modern, more economical, and will be far easier to defend. It will be free of complex, costly, vulnerable locks and seaways. It will serve the future as the Panama Canal we know has served the past and the present." Thanks to the $17 million Congress had authorized to study four possible routes—two in Panama, one in Colombia, and one through Nicaragua that might also include part of Costa Rica—the United States would be well-equipped to modernize maritime transportation. He ended the speech on a forward-looking, anti-imperialist note: "The age before us is an age of larger, faster ships. It is an age of friendly partnership among the nations concerned with the traffic between the oceans. This new age requires new arrangements."[58]

The strategy could not have worked better. Despite having made no reference to PNEs, U.S. news coverage highlighted the flashy concept of a Panatomic Canal, which deflected attention from the unpopular idea of relinquishing U.S. sovereignty over the Canal Zone.[59] In Panama, the announcement elicited "chagrin" among nationalists, as the *New York Times* reported.[60] Mann likewise soon reported to Bundy that the president's statement had thrown the Communists off balance. As for the upcoming Flag Riots events, the Canal Zone and Panamanian governments had agreed to lower the flags of the two nations to half-mast, and President Robles had assured the U.S. ambassador that no trouble would be tolerated.[61]

On January 9, 1965, a huge contingent of Panamanian national guardsmen lined the boulevard separating the Canal Zone from Panama City, and approximately two thousand young Panamanians marched to the gravesite of the first student killed a year earlier. Later on, after demonstrators raised banners and chanted against both the U.S. and Robles governments, guardsmen deployed tear gas grenades against a group of about two hundred.[62] Otherwise the day that would eventually be commemorated as Día de los Mártires (Martyrs' Day) passed with neither fatalities nor negative publicity, just as Johnson had hoped.

Conclusion

The Flag Riots helped transform the nuclear seaway idea from a U.S. presidential delaying tactic into a serious diplomatic instrument for resolving the persistent tensions caused by the 1903 treaty between the United States and Panama. The sea-level canal's economic, defense, and political rationales each proved insufficient to justify the project.[63] But the unprecedented scale of anti-U.S. violence in January 1964 accentuated the value for U.S. decision makers of merging the three factors. Contemporaneous technical advances in PNEs further strengthened the case for committing significant scientific and diplomatic attention to the several possible routes the megaproject might take.

Ernest Graves, one of the Army Corps engineers who worked with the Plowshare physicists at Livermore on the 1964 canal study report, later acknowledged the sea-level canal as a technological solution to a multitude of problems: "We were going to build a sea-level canal which didn't require a big American community, and it was going to have unlimited capacity so that we would never have to worry any more about whether there was enough water [during droughts] or how long it took to lock a ship through. All that would go away."[64] But the nuclear seaway was no ordinary technological fix. An immense amount of scientific, engineering, and diplomatic groundwork remained before the problems produced by the 1903 treaty could ever go away.

It is true that, like his predecessors, President Johnson used the nuclear sea-level canal idea to his advantage. However, framing the proposal as a cynical ploy or high-modernist boondoggle limits our understanding of how it functioned as a complex, qualified instrument of technological anti-imperialism. While it did serve to buy time and distract angry stakeholders, that does not mean the proposal was fictitious. Whether or not he believed it would ever come to fruition, Johnson's advocacy of the nuclear canal set in motion a process that would have important, unpredictable side effects pertaining to politics, science, technology, and the environment.

Nor was the proposal particularly peculiar. In fact, the nuclear seaway resembled other Johnson administration initiatives to mobilize science and technology for diplomatic ends in the so-called Third World, from weather modification in India and Pakistan to herbicidal warfare in Vietnam. Moreover, the Panatomic Canal embodied the broader Cold War ethos of faith in science and technology that "encouraged bold environmental interventions" to reshape and optimize the landscape for human use.[65] But while a sea-level canal would facilitate the abolishment of the colonialistic Canal Zone, how the costs and benefits of blasting a new waterway with thermonuclear bombs could be adequately assessed, let alone distributed equally, remained one of the many critical questions facing the proposal's proponents.

PART II

The Panatomic Canal

CHAPTER 4

Navigating High Modernism

PART II EXPLORES THE WORK of a key set of actors in the nuclear sea-level canal story who have received a bad rap: the members of the Atlantic-Pacific Interoceanic Canal Study Commission (CSC), whom President Johnson appointed and Congress authorized to determine the feasibility of constructing a new Central American seaway. Ever since the commission released its final report in 1970 recommending conventional rather than nuclear construction of a new channel to the west of the original one, it has been criticized for wasting millions of dollars, minimizing ecological risks, and rubber-stamping a foregone conclusion.[1]

But far from failing, the CSC largely fulfilled what Johnson deemed its "difficult and complicated mission" of rendering a responsible recommendation despite the many constraints and uncertainties it faced.[2] Taking a closer look at the group complements the insights of Plowshare historians on the hubris of the Livermore Laboratory nuclear scientists and engineers driving the project.[3] Unlike Plowshare's strident proponents, the canal study commissioners kept a low profile and provided a semblance of objectivity, even as they self-consciously committed the majority of their funds to investigating the remote routes deemed most amenable to nuclear excavation. But the process wound up being far from straightforward; as one member concluded after five intense years, "We proved the opposite of what we expected."[4]

The commissioners were not scientists, but some had engineering expertise, and all were of the generation born around the turn of the century that witnessed world-changing inventions—automobiles, penicillin, plastic, synthetic pesticides, jet airplanes, and nuclear weapons, to name a few. Such technoscientific developments promoted an uncritical acceptance of perpetual progress and inexorable movement forward no matter the negative consequences for landscapes or "people in the way" of modernity.[5]

The commissioners embraced elements of high modernism, the early-to-mid-twentieth-century ideology that invested great faith in the ability of science and technology to improve humanity via state-sponsored development projects.[6] Full high-modernist development only occurred in totalitarian states like the Soviet Union and People's Republic of China, where forced collective agriculture constituted a major application of the philosophy. In the capitalist West, democratic institutions vitiated the ability of planners to impose their absolute visions of order upon the land and people. But North American megaprojects such as hydroelectric dams and urban renewal did feature top-down planning and other high-modernist hallmarks.

The CSC operated in the context of a presidential administration that also privileged technological solutions to political problems and disdained criticism thereof, as exemplified by the near-simultaneous escalation of chemical herbicidal warfare in Vietnam. Indeed, the major architects of the war in Vietnam were the greatest proponents of the nuclear sea-level canal, or at least of renegotiating relations with Panama in relation to it. After the Flag Riots, Secretary of Defense Robert McNamara and National Security Advisor McGeorge Bundy promoted the announcement of the historic foreign policy change regarding the U.S.-Panama treaty in 1964. Later, Bundy's successor Walt Rostow likewise advised, "Since the Panama Canal is a rapidly wasting asset, we must plan beyond its life expectancy, decide upon a suitable replacement, and formulate policies to keep the Isthmian region within the Inter-American system and susceptible to US influence." The sea-level canal held the key to modernizing the technological and political Panama Canal complex.[7]

Under such circumstances, it was not at all inevitable that the commission would advise against nuclear construction of a new canal. Their work in getting to that point sheds light on the evolution of environmental impact assessment during the pivotal years between Project Chariot—the Alaskan nuclear harbor affair—and the enactment of the National Environmental Policy Act (NEPA).

Congress passed NEPA in December 1969, after several high-profile cases of pollution focused public attention on the negative environmental consequences of modern industrial society. NEPA sought to instill environmental values into federal megaproject decision-making by requiring federal agencies to conduct preliminary environmental impact studies. The law also provided opportunities for public input and empowered citizens to challenge technocratic projects via the courts. Such developments traced in part to the revolutionary argument of Rachel Carson's best-selling 1962 book *Silent Spring*—that "the authoritarian temporarily entrusted with power" had no right to make unilateral decisions regarding the use of technologies affecting society.[8]

The commissioners and their executive director showed no awareness of Carson's book, and they derided environmentalism as a passing fad. However, pressure from marine scientists forced them to consider how broadly they should define the scope of the environmental data needed to determine the sea-level canal's engineering feasibility. Was it enough to fund limited studies on how radionuclides released by peaceful nuclear explosives would affect local food chains and ecosystem cycles? Or should the ecological studies also encompass the adjacent oceans, to address the potential nonradioecological effects of seaway construction? The latter question dogged the commission throughout its existence but especially during the second half of its tenure, from early 1968 to December 1970.

This chapter examines how the CSC (also known as the Anderson Commission, after its chair, Robert B. Anderson) and its consultants engaged in multiple sets of negotiations in the service of the nuclear sea-level canal endeavor during its first two-and-a-half years, from the spring of 1965 to late 1967. By divulging the twists and turns of their deliberations, the commission's scrapbooks and declassified meeting transcripts shine light on an evolving approach to environmental impact assessment in the pre-NEPA era.

The concept of "negotiated high modernism," which refers to the consultations and compromises that democratic governments must engage in to bring unpopular megaprojects to fruition, helps us understand how the CSC coped with the challenges of rendering a responsible decision.[9] The commissioners had to deal not only with routine impediments, like short deadlines and inadequate funds, but also with daunting diplomatic and physical constraints: the U.S.-Panama treaty talks (for which Anderson also served as the lead U.S. negotiator), the Johnson administration's hesitancy about permitting excavation experiments at the Nevada Test Site, unexpected opposition to the field surveys in Panama and Colombia, and the short climatological windows during which research could be conducted each year in Nevada and the Darién.

The commission navigated those rough waters with a degree of flexibility that was remarkable for large-scale technocratic planners of the era. To take but three contemporary examples, the architects and implementers of the Aswan Dam, midcentury Manhattan, and the chemical war in Vietnam demonstrated little if any consideration of the ecological and social toll.[10] The members of President Johnson's canal study group deserve more attention for how they reconciled their faith in the promise of nuclear excavation with external demands to address its nonfinancial costs.[11] A significant body of evidence supports the argument that "the people who were planning Plowshare were choosing to ignore those negative consequences."[12] It is equally important to distinguish Plowshare's uncompromising idealists from those tasked with providing a realistic assessment of its signature project.

Engineering the Commission

At the behest of the Johnson administration, Congress established the Atlantic-Pacific Interoceanic Canal Study Commission near the end of the 1964 session, eight months after the Flag Riots exposed the political unsustainability of the U.S.-run Panama Canal Zone. The law authorized the commander-in-chief to appoint "five men from private life" to conduct an investigation of enormous scope to determine where and how to build an isthmian sea-level canal. To oversee the collection and analysis of data pertaining to national defense, foreign relations, intercoastal and interoceanic shipping, and engineering feasibility, the law permitted the commission to draw on any federal agency and to spend up to $17.5 million.[13] While the deadline was tight—June 30, 1968—the budget was not inconsiderable. By means of comparison, for example, Congress allocated $9 million for the 1965 fiscal year to the Arms Control and Disarmament Agency, a federal bureau established in 1961 to prevent the spread of nuclear weapons.[14]

Months later, Representative Dan Flood was still fuming over the law and Johnson's surprise announcement about renegotiating the Hay–Bunau-Varilla Treaty with Panama. The Democratic Pennsylvania congressman had been defending the sovereignty of the United States over the Canal Zone since the 1958 riots and had worked to ensure that the CSC members would not be the government officials desired by Johnson. However, the president's delay in making the appointments, and press coverage that the cabinet officials Stephen Ailes and Thomas Mann had visited Nicaragua, Costa Rica, Colombia, and Panama in January 1965 to discuss canal options, led Flood to allege a cover-up.[15] Convinced the commission would approve the Panama Canal Company's 1947 recommendation to convert the present waterway to sea level, and that the nuclear routes were propagandistic distractions, Flood reminded his colleagues of why his hero Theodore Roosevelt had rejected the sea-level arguments back in 1906. He also eviscerated Johnson's diplomatic overture on the grounds that "wresting control of the Panama Canal from the United States and its internationalization have been Red objectives since 1917." Flood even praised the defiant students who sparked the deadly 1964 riots by raising the Stars and Stripes: "I would prefer to have children from our American high schools to formulate our canal policies rather than hidden appeasers and sappers in the executive departments."[16] Flood validated his status as the "all-time nut on the subject of the Panama Canal," as one of Johnson's aides later described him, by demanding that Ailes investigate the fiftieth-anniversary commemorative catalog published by the Canal Zone Library-Museum. Because it referenced only a few of his own speeches, Flood denounced it as pro-seaway "bibliographic sabotage."[17]

Flood's antics made for a persistent thorn in the CSC's side. When, for example, Chairman Anderson declined Secretary of the Army Ailes's offer to serve as a consultant, a colleague explained, "I assume this sensitivity about Pentagon influence stems from Dan Flood's tirades about the whole thing being a rubber stamp operation to approve a foregone position that Ailes and Mann sold to the President."[18] When the commissioners objected to having their photos and detailed résumés included in their second annual report, the executive secretary responded, "May I offer the excuse that a number of people have criticized the Commission as not having competence for their jobs. I thought I had better put your background in there to show your competence—[to neutralize criticism from] Dan Flood, primarily."[19] Most significantly, later efforts to amend the authorizing legislation to provide additional funds and time had to be crafted with sympathetic members of the House committee overseeing the Panama Canal (Merchant Marine and Fisheries) so as not to attract Flood's ire. At one pivotal point, Representative Leonor Sullivan confided that she would not be able to slip a requested change through by simply listing it on the committee calendar. "This time she fully expects it is going to be argued on the floor and Dan Flood is going to be in full bloom."[20]

Flood's diatribes, combined with the intensifying war in Vietnam, probably contributed to President Johnson's seven-month delay in appointing the canal study commissioners.[21] To keep the treaty negotiations and seaway feasibility studies interlinked, Johnson selected Robert B. Anderson, the special representative for U.S.-Panama relations, as CSC chair. The other members had a variety of distinguished backgrounds. Serving as vice chair was Robert G. Storey, a Nuremberg prosecutor who had since founded a legal foundation at Southern Methodist University. Milton S. Eisenhower, the president of Johns Hopkins University, had directed a commission on U.S.–Latin American relations during his brother Dwight's administration; the resulting 1963 book predicted the escalation of anti-U.S. violence in Panama.[22] Raymond A. Hill was a renowned water resources development expert and lead author of a 1938 compact that addressed long-standing water rights disputes over the Rio Grande.[23] And finally, retired Brigadier General Kenneth E. Fields had commanded a famous World War II engineer combat group, assisted General Leslie Groves of the Manhattan Project, and served in the Atomic Energy Commission.[24] Except for Eisenhower, the men possessed the kinds of expertise envisioned by Flood for his ideal committee of independent assessors, but the congressman still tried for months to have the "legislative monstrosity" establishing the commission repealed.[25]

Performing most of the commission's day-to-day work, which included coordinating with the many subcommittee chairs employed by other federal agencies, communicating with members of Congress and the press, and drafting the annual reports, was Colonel John Sheffey. Having worked in Panama since 1961 as the military assistant for canal affairs to the secretary of the army, he was well versed in isthmian politics. Earlier in his career, he had also completed a three-year program in nuclear energy and weapons. Sheffey, then in his midforties, retired from the military to take on what he considered a prestigious assignment as the study commission's executive director.[26] Decades later he attributed his enthusiasm for the job to Plowshare's two most outspoken spokesmen, the Livermore physicists Edward Teller and Gerald Johnson: "I changed my whole life because I believed them, and I believed that the greatest thing in the world for me [was] to be a part of that first great nuclear construction project."[27]

The Anderson Commission held thirty meetings during its five-year existence (it received two congressional extensions, the final one until 1970). On occasion, high-level cabinet members attended, but the typical meeting featured presentations by representatives of one or more of the federal agencies responsible for the five study subgroups: foreign policy (State Department), national defense (Defense Department), shipping (Transportation Department), canal finance (Treasury Department), and most important for our purposes, engineering feasibility (U.S. Army Corps of Engineers).

The postwar mandate of the Army Corps of Engineers emphasized flood control, navigation works, and other aspects of water resources development, but the agency retained strong links to the atomic energy establishment. The corps had played a key role in organizing the Manhattan Project, and in 1962 established the Nuclear Cratering Group at the Lawrence Livermore Laboratory. Like the Livermore Plowshare physicists—and civil engineers more generally—corps personnel sought to reshape the landscape for utilitarian purposes. Due to their shared values, and the corps's influence in Congress (a function of its pork barrel water projects), the Plowshare-Corps partnership was mutually beneficial.[28] Accordingly, the CSC designated an Army Corps officer as its official engineering agent. The agent attended almost every meeting, and three men fulfilled the role over the commission's life span (Harry G. Woodbury, Charles C. Noble, and

Richard S. Groves).[29] To provide additional updates on the engineering feasibility studies, other frequent guests were John S. Kelly (the director of the AEC's Division of Peaceful Nuclear Explosives) and fellow AEC officials, who subcontracted the canal studies to both academic and private research organizations.

The canal study commissioners had much of their work cut out for them because the Livermore physicists and Nuclear Cratering Group engineers had been working on the sea-level canal project for years. As revealed at the most recent Plowshare symposium in 1964, they had focused on the two shortest, least-populated isthmian routes, both of which traversed the Darién: one, known as Route 17 or the Sasardi-Morti route, through Panama's dense eastern forests bounded by the Sasardi and Morti Rivers, and the other, known as Route 25 or the Atrato-Truando site, through Colombia's marshy Atrato River valley (map 1.2). Yet good maps, let alone subsurface geophysical data, remained elusive. More than 150 years after Humboldt had heralded the vast region's potential for an interoceanic communication, even the exact height of the Continental Divide along Route 17 remained unknown. But the estimate of 1,100 feet above sea level posed an exhilarating challenge, as expressed by Nuclear Cratering Group leader Ernest Graves: "A cut this deep by any means would be an engineering achievement of the first magnitude. To do it in less than a minute with a single explosion staggers the imagination. Nevertheless, the scientists and engineers who have studied the problem have faith it can be done."[30] The construction of the Panama Canal, Graves reminded his audience, had been equally astonishing five decades earlier.

Because the atomic seaway would take ten to thirteen years to complete, Seaborg, Johnson, and Kelly testified before Congress in January 1965 that the field surveys and nuclear cratering tests should start as soon as possible.[31] Yet the Anderson Commission did not begin meeting until the late spring, by which time it was already behind schedule. Due to the original three-year congressional limit and the short tropical dry season, which lasted from December/January to April, Corps of Engineers representatives had outlined an ambitious schedule of data collection and site surveys. Extensive supporting infrastructure—weather stations, field offices, camps, supply points, and roads—would have to be built to accommodate hundreds of workers responsible for collecting two kinds of data during the first two dry seasons of 1965–66 and 1966–67: (a) topographic surveys and geological, hydrological, and hydrographic studies to provide basic information about the drainage areas, sedimentation processes, coastlines, and seafloor along each of the routes in Panama and Colombia, and (b) more specific meteorological, air blast, seismic, and bioenvironmental data to assess the radiological safety of nuclear excavation. The final year of 1967–68 would be reserved for evaluating all the data to determine the most feasible, cost-effective channel designs, as well as the projected schedule of nuclear detonations and area evacuations.[32] Meanwhile, if all went according to plan, Plowshare scientists and technicians would be conducting six experiments at the Nevada Test Site to see how various configurations of PNEs operated in nature rather than in theory.

Uncharted Territory

Not only was time not on their side but also, from the start, the commission members harbored deep concerns about the costs posed by the field studies. The infrastructural costs alone—constructing the camps and data collection stations, clearing center line trails across the isthmus, and providing communications and medical support along the two routes—would consume $2 million of the $17.5 million budget.[33] Ideally, the equipment would be set up prior to the dry season of January 1, 1966, but Congress resisted releasing funds before the survey agreements with Panama and Colombia had been inked, making for yet more delays and logistical headaches.[34]

Another worrisome constraint over which the commission had no control was the test ban treaty restriction against depositing radioactive debris in adjacent nations. Despite the commissioners' enthusiasm for PNEs, they knew there would be no point in conducting any sea-level canal studies if the administration had no intention of spending political capital to amend the protocol to allow peaceful nuclear experiments. Yet Plowshare's unresolved relationship to the Limited Nuclear Test Ban Treaty did not discourage the AEC. At the second meeting as well as later ones, Kelly maintained that the Russian language text of the treaty provided for a more liberal interpretation of the ban on radiation outside national borders and that every test shot would release some radiation—which would not pose a serious health risk anyway.[35]

Other federal agencies, especially the State Department and Arms Control and Disarmament Agency, viewed Plowshare as a threat to nuclear weapons nonproliferation initiatives. The third CSC meeting, in July 1965, featured a heated discussion among representatives of the AEC and the Arms Control and Disarmament Agency about the potential of Plowshare experiments to cause an international incident by venting radiation across the border. The impasse seemed intractable; while President Johnson wanted the U.S.-Panama treaty negotiations to wrap up soon, the sea-level canal treaty hinged on the feasibility of nuclear engineering, which required experimental explosions at the Nevada Test Site.[36]

The scope of the engineering feasibility studies also occupied the agenda of the early meetings. From the start, the CSC and its AEC partners recognized the importance of researching isthmian food chains and ecosystems to determine how their human users would be affected by the radiation released by PNEs. Even as the AEC assured the public that radioactive fallout carried minimal health risks, the agency provided a major source of support for ecosystem ecologists during the Cold War.[37]

But might other kinds of bioenvironmental research also be needed to provide a yardstick against which to measure the changes caused by seaway construction? That query came from an unexpected source, Chairman Anderson's deputy treaty negotiator. John N. Irwin II, a fellow Republican and Manhattan-based lawyer, attended the CSC meetings when his boss's busy schedule kept him away.[38] Irwin's job was to brief the commission on the latest developments in the U.S.-Panama treaty talks, but he also bugged them about an interest that occupied his leisure time. As a trustee of the New York Zoological Society, Irwin mingled with elite conservationists, such as Laurance Rockefeller, as well as scientific employees of the Bronx Zoo.[39] One of the zoologists asked him a question that he in turn posed to the commission in July 1965: Would the data collection efforts along the two Central American routes also seek to elucidate the non-radiation-oriented effects of a sea-level canal on marine life, and might the Smithsonian Institution take part in such a study?[40]

The zoologist had likely read the latest issue of *Natural History* magazine, which contained an article titled "Mixing Oceans and Species" by an up-and-coming marine biologist at the Smithsonian's Panama research facility, which occupied an island in the drowned Chagres River valley, the reservoir of the canal. The essay addressed the "interesting biological problems" regarding the marine consequences of building a sea-level canal. Unlike the existing lock canal, which contained a large freshwater reservoir that prevented most marine species from transiting, a sea-level channel would join the Atlantic and Pacific Oceans for the first time since the rise of the isthmian land bridge during the late Pliocene. Accordingly, the author, Ira Rubinoff, speculated on the evolutionary and ecological effects of intermixing the Atlantic and Pacific Oceans.[41]

Engineering Agent Harry Woodbury dismissed Irwin's query, stating that many organizations sought to participate in the sea-level canal studies on aspects that fell "far beyond the scope which is of concern to the Commission." He conceded that the corps had a history of working with the Smithsonian on archaeological issues raised by construction projects. But the Smithsonian's proposed biological baseline survey surpassed the essential biological questions of seaway construction that did not involve radioactive hazards.[42]

When the two commissioners with engineering backgrounds likewise called for drawing a sharp distinction between desirable and essential data, Irwin provided a friendly word of warning: "I bring it up so that the Commission will know what will be in the minds of ecologists, zoologists, and others. You may or may not at one time consider whether or not you want to broaden this scope, not from the pure feasibility point of view, but from the point of view of being able to answer people on the effect."[43] Perhaps recalling what had happened with the Alaskan harbor proposal, Irwin again used his time at later meetings to caution that a narrow bioenvironmental study might generate criticism from scientific groups, even though "they may not be significant in the sense of popular reaction."[44]

A Republican diplomat was thus the first nonscientist to give the Democratic appointed presidential commission a heads-up about the importance of paying more attention to broad-scale ecological assessment. Fifteen years later that would have seemed strange, but for the first seven decades of the twentieth century, moderate Republicans supported many facets of protoenvironmentalism, from wilderness preservation to utilitarian conservation to population planning, often in close concert with scientists.[45]

Knowing he was outgunned, Irwin conceded it would be sufficient for the commission to invite other agencies, such as the National Academy of Sciences, to contribute to a nonnuclear ecological assessment using their own funds. The group agreed, but otherwise did not understand Irwin's concern. After all, the private research organization that had won the AEC's bioenvironmental contract, the Battelle Memorial Institute, planned to collect terrestrial baseline data as well as information on oceanographic currents, temperature gradients, and marine life on either side of the isthmus. As Kelly explained, "I think our bioenvironmental program while it is principally addressed to preventing radioactivity getting to man, in tracing it from the time it is released by the explosive until the time it gets to man . . . will develop an awful lot of this ecological information you were talking about; and this information would be available for people to evaluate." Besides, additional research could always be conducted later if the government decided to proceed with construction.[46]

When the discussion turned to another major concern of the Anderson Commission, the managing of public relations, Kelly invoked the infamous Project Chariot to draw a different lesson than that suggested by Irwin: "We don't advocate a grandiose program of selling nuclear explosives, but I think we should be in the position of taking the initiative of explaining what we are doing." When the Chariot project began, he explained, an agreement between the State Department and AEC had precluded the latter from taking a proactive stance, "and we got into trouble because that is all we could do, answer questions. People don't want to detract from projects, but they never asked the questions the answers to which were meaningful."[47] By contrast, when the AEC orchestrated press coverage of the shots at the Nevada Test Site where they "were not limited to the requirement of just responding to inquiries" and could instead "take some actions to explain what [they] were doing," public trust remained high.

Other members of the atomic energy establishment echoed Kelly's attitude about the proper way to mold public opinion. As the former AEC commissioner and Nobel prizewinning physical chemist Willard Libby told a journalist in 1966, both of Alaska's senators had supported Project Chariot and thus the plan should have proceeded. "But our overcautious preparations created a public-relations problem. If the test was so safe [people asked], why did the AEC spend $3 million to count all the birds and animals in the area? Our cautiousness gave the lie to our reassurances about fallout, and ruined the project. If we'd been that careful about using the open-hearth furnace, we wouldn't be making steel today."[48] Libby's interpretation overlooked the intense public opposition that had led the AEC to expand the scope of the Chariot feasibility studies (rather than the other way around), but it aligned with the AEC's dismissiveness regarding public fears of radiation.[49]

The Thing That Makes the Inevitable Come to Pass

Unlike Plowshare's assertive proponents, the Anderson Commission members stayed out of the spotlight, especially as they finalized plans to spend most of their $17.5 million on the two sea-level canal sites deemed most amenable to nuclear excavation, the Darién portions of Panama and Colombia.[50] The other two routes Johnson had identified in his December 1964 announcement remained on the back burner. Despite Flood's allegation that the commission would recommend converting the existing canal to sea level, as the Panama Canal Company had advised in 1947, the CSC invested little in that option. As for the Nicaragua–Costa Rica route, the group did fund the Inter-American Geodetic Survey to produce the first modern topographical maps of the area, but otherwise did not seriously consider it since, among other reasons, a seaway would drain Lake Nicaragua.[51]

The press release for the commission's fact-finding trip to Panama in August 1965 proclaimed its impartiality: "Our Commission is only beginning its work. We begin without preconceptions."[52] But behind closed doors, the commissioners grappled with the preconceptions that permeated their mission. At their seventh meeting, in November 1965, they grilled National Security Advisor McGeorge Bundy about the administration's commitment to the atomic seaway. Commissioner Hill, who possessed the most experience in the contentious realm of water resources management, pressed Bundy about whether the sea-level canal decision would ultimately come down to economics or politics. Bundy hedged that a new waterway appeared to be on the horizon due to its technological and diplomatic superiority over the outdated existing channel, but stated that it was by no means a fait accompli. Yet when pushed further on the prospects of such an expensive megaproject, Bundy conceded, "I think you will find you are in a more realistic position if you assume that this is something that is going to happen."[53]

Hill kept up his cross-examination, stating that prior reports had portrayed the seaway with an air of inevitability. "Let me say there is just a shade of that in the whole Plowshare exercise, too," replied Bundy, "But the converse of that, Mr. Hill is that the thing that makes the inevitable come to pass is effort." The chair ordered the ensuing discussion off the record, a sign of the topic's immense sensitivity. Later, Bundy addressed one last question on the record, about how the commission should interpret the word *feasibility* as used in the authorizing legislation. "If we needed to dig a new canal and have to get it done by the end of 1968, we could do it," he snapped. "Ergo, in the strict sense we already know that a sea-level canal is feasible." That did not of course preclude a thorough assessment of costs "and all the other practical considerations that belong in a recommendation to the Government of the United States." On his way out the door, Bundy reminded the commission of who was really in charge: "Obviously if we wanted it enough *today*, we could afford it, and we could drive it through."[54]

Yet despite Bundy's high-modernist mic drop, the ability of the world's wealthiest nation to afford grand projects was eroding as the administration committed more and more resources to the war in Vietnam. As the president announced two months later, "The budget for 1967 bears the strong imprint of the troubled world we live in."[55] Several agencies faced major cuts; the AEC lost $103 million. Nonetheless, Seaborg insisted at the January 1966 CSC meeting that by consolidating the remaining nuclear excavation experiments, Plowshare personnel could answer the feasibility question by the June 30, 1968 deadline.[56] Commissioner Hill took such optimism with a grain of salt: "I don't think anybody here can deceive himself that you are going to know about nuclear excavation in time to complete a canal by 1980," the earliest possible date of operation.[57]

Not only were budgets being cut left and right, political and meteorological impediments loomed large. The AEC had planned to conduct six shots starting with the Cabriolet experiment at the Nevada Test Site that spring. The timing of each test was essential to avoid potential releases of radioactive fallout during the grazing season. But Johnson postponed the Cabriolet shot during the critical 1966 window for fear of violating the Limited Nuclear Test Ban Treaty and disrupting the current Soviet-U.S. nonproliferation discussions.[58]

Thousands of miles south of the Nevada desert, the Darién fieldwork also failed to start as planned. Gaining permission from the governments of Panama and Colombia proved more difficult than expected because legislators in both nations perceived the field studies as diversionary tactics. Panamanians sought to keep the diplomatic focus on abrogating the 1903 treaty, while Colombians objected to being used as leverage against Panama.[59] Not until February 15, 1966, did the State Department secure a site survey agreement for the Route 17 studies in Panama, and the Route 25 negotiations with Colombia did not conclude until October 25, 1966. The late start cost a great deal of time and money, as did the unexpected allocation to Southeast Asia of military helicopters and other equipment needed by the Darién surveyors. Barely a year after its formation, the Anderson Commission was behind schedule and begging Congress for more funds.[60]

Except for the burning of a single U.S. flag, the second anniversary of the Flag Riots in January 1966 passed without incident in Panama.[61] But Johnson had little to celebrate, as the treaty negotiations dragged on. Moreover, geopolitical, economic, and meteorological constraints were converging so as to subvert the technocratic "air of inevitability" and the assumption underlying the Atlantic-Pacific Interoceanic Canal Study Commission—that an engineering solution to the multifaceted problems posed by the lock canal could be achieved by mobilizing science and technology to reshape the political and hydrological geography of the isthmus. By the time the engineering feasibility field studies finally took place from 1966 through 1969, the commissioners found themselves having to adapt to obstacles that decelerated the institutional momentum underlying Plowshare.

Surveying the Space Age Jungle

Humboldt would have been stunned to learn how rudimentary the scientific knowledge of the Darién remained so many decades after he had called for it to be "levelled." Although Panama's Barro Colorado Island, in the middle of the canal's drowned Chagres River valley, had become a premier site for tropical biological research, few researchers ventured east to the Darién. Its lack of infrastructure still made overland travel an ordeal, and thus the sporadic efforts of outsiders to drive the 310-mile-long gap in the Pan-American Highway between Chepo, Panama, and Quibdó, Colombia, generated widespread interest. As one explorer marveled, "even in today's space age, man still has frontiers to cross in forgotten corners of his Earth."[62]

In the 1960s, thick rainforests shrouded the Continental Divide's mountain ranges and the lowlands of eastern Panama, and vast marshlands permeated the Atrato River valley of northwestern Colombia (the Darién biogeographic region encompasses both countries, but the political province of Darién is confined to Panama). Eastern Panama's population of approximately fifty thousand consisted of Indigenous tribes, African descendants, and mestizo settlers. The Kuna, or Guna, famed for their independence and mola artwork, numbered twenty-one thousand and practiced subsistence fishing and slash-and-burn farming in the upper Chucunaque River valley and on the San Blas coast, near the Caribbean end of the proposed canal route. The fifteen hundred Indigenous Chocó (now known as the Emberá and Wounaan) occupied villages along rivers draining into the Gulf of Miguel, the Pacific terminus of Route 17.[63]

By the time the U.S. and Panama worked out the Route 17 site survey agreement in mid-February 1966, the dry season was well underway, and thus the corps's Canal Zone–based Office of Interoceanic Canal Studies scrambled to build roads and base camps before the torrid humidity, fog, and rains returned. Colonel Alexander G. Sutton Jr. had taken charge of the office the previous summer, having spent four years directing the corps's Waterways Experiment Station in Vicksburg, Mississippi.[64] The facility's large-scale models of the Mississippi and other rivers helped engineers predict the effects of flood-control structures and otherwise bridge the realms of hydraulic science and engineering.[65] Yet despite their sophistication and utility, models constitute only abstract representations of nature. Taking the measure of the formidable Darién would entail very different kinds of considerations and actions than operating a control panel of knobs and switches.

The work did not begin well. The Johnson administration's month-long delay in announcing the site survey agreement led to misunderstandings, as did a lack of transparency on the ground. The *Panama American*, a newspaper that primarily served U.S. residents of the Canal Zone, reported in mid-March that eyewitnesses had viewed a few dozen men setting up tidal and weather stations, as well as a coast-to-coast surveying track, across the eastern Darién. Yet when reached for comment, the U.S. embassy and other offices denied that the Route 17 work had begun. The author considered such secrecy pointless, given that land

FIGURE 4.1. Equipment delivered via U.S. Navy tank landing ship to construct a meteorological station on Soskatupu Island near the Atlantic terminus of the proposed Panamanian nuclear route (Route 17), ca. early 1966. Note the Guna canoes at lower right. The original caption includes the statement, "Cuni [sic] Indians came from many miles in their dugout canoes to watch the activity." APICSC Scrapbook, RG 220, U.S. National Archives and Records Administration, College Park, Md.

speculators had long since obtained what they could along the well-publicized routes. Two days later, another embarrassing article reported that Guna and Chocó delegates had traveled to Panama City to protest the unloading of heavy equipment on the north and south coasts without their consent (fig. 4.1).[66]

Behind the scenes, the Office of Interoceanic Canal Studies technical liaison staff blamed the negative coverage on efforts to "interweave anti-canal study propaganda with the plight of these Indians," as well as domestic Panamanian opposition to Foreign Minister Fernando Eleta, the leader of Panama's treaty negotiating team. Eleta had provoked anger among his fellow citizens by not submitting the site survey agreement for advance approval to the National Assembly and by waiting until April to confirm that the field studies had actually begun.[67] He might have been concealing a conflict of interest; according to a confidential document in the Anderson Commission's files, Eleta supported the Route 17 proposal to stimulate development of the area, where he owned property. The business leader, who held an undergraduate degree in structural engineering from MIT, also asked that an upcoming *Atoms in Action* exhibition scheduled for Panama focus more on the promise of nuclear excavation.[68] Latin American opposition to nuclear weapons proliferation, and to the atmospheric nuclear tests France initiated in the South Pacific in 1966, would necessitate major outreach efforts to achieve buy-in for the atomic waterway.[69]

Convincing Panamanians to accept PNEs would be one thing; in the meantime, the commission learned important lessons from the rush to set up the field studies infrastructure. The fruits of establishing strategic media contacts appeared in late May 1966, when the *Panama American* published a lengthy article in English and Spanish titled "What's Happening in Darien Survey?"[70] The paper's editor had asked the local corps office numerous times for permission to visit the site with a photographer, and only received it after assuring that no unfavorable coverage of the project's political dimensions would appear. As the Office of Interoceanic Canal Studies liaison officer boasted to the commissioners back in Washington, the article's flattering approach made it more of a press release than a journalistic exposé. The corps even sent extra copies as a morale-building effort to the three dozen men in the field, who were experiencing grueling heat and the predations of biting mosquitoes and vampire bats.[71] The upbeat news story got picked up by international media, which facilitated the commission's task of controlling the narrative that the field studies were at last underway and under control.[72]

Another hard lesson entailed paying more respect to residents of the survey areas, especially the semiautonomous Guna, a people with a long history of resisting Spanish, Panamanian, and U.S. domination.[73] At the June 1966 CSC meeting, Engineering Agent Woodbury reported that Sutton's team had finally attained permission from the Guna leadership to proceed, thanks to "considerable help from the Panamanian representatives who sent a lady with us into Kuna country along with a Panamanian doctor to lay the groundwork for this."[74] The lady, Reina Torres de Araúz, would become Panama's most legendary anthropologist before her untimely death at age forty-nine in 1982. While still in her twenties, she had become a professor of anthropology at the University of Panama and participated in the Trans-Darién Expedition, an effort initiated by two Canadians to cross the gap via Land Rover station wagon. She and her husband, cartographer Amado Araúz, joined them in February 1960, spending 134 arduous days cutting trails and building bridges and rafts to drive (or float) to where the highway resumed in Colombia.[75] Along the way, she conducted crucial research on Indigenous cultures.

Torres helped the corps broker an agreement by which the U.S. would provide medical assistance and compensation for damaged Guna trees and gardens. The "What's Happening in the Darien Survey" article had noted that the coast-to-coast surveying program required "cut[ting] down thousands of trees, most of a useless nature." Yet the trees were not useless to the Guna, who negotiated reimbursements of $2 to $5 for palms and $7 for avocado trees cleared

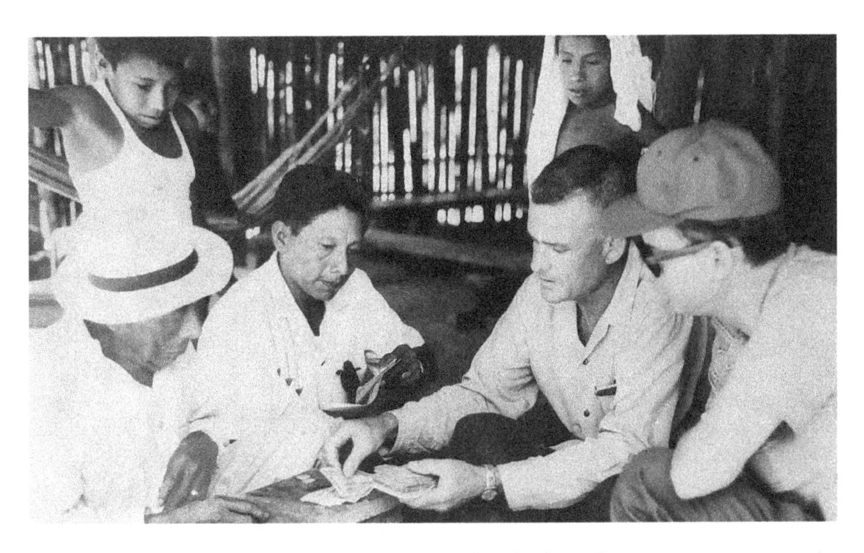

FIGURE 4.2. U.S. Army Corps of Engineers colonel Alexander Sutton paying a stack of U.S. bills to Guna chief Yabiliquina to compensate for trees and vegetation removed during the construction of a weather-recording station in the Darién, Soskatupu, Comaro de San Blas, September 20, 1966. APICSC Scrapbook, RG 220, U.S. National Archives and Records Administration, College Park, Md.

for the scientific facilities and center line (figs. 4.2 and 4.3).[76] Through 1968, Torres directed several AEC-funded studies on the human ecology of Panama's eastern residents, who would be most affected by canal construction and radioactive contamination of their food chains.[77] Her results later informed the U.S. Department of Transportation environmental impact statement requirements for the proposed completion of the Pan-American Highway through the Darién Gap.[78]

By the 1967 dry season, the Anderson Commission was on firmer ground. Congress had extended its deadline by a year to June 30, 1969. Along Route 17, all sixteen hydrology stations, including tidal gauge, rain gauge, and river gauge stations, were up and running. About 100 U.S. citizens and 250 Panamanians were collecting hydrographical, meteorological, seismic, biological, and medico-ecological data, as reported in a favorable *New York Times* article.[79] Coverage by local journalists emphasized the benefits to Panama, which included both jobs and valuable data about potential mineral, hydroelectric, agricultural, and fishery resources.[80] The field studies to the southeast in the Atrato River valley of Colombia had also at long last begun. Back in Washington, the commission's various working groups were completing their initial drafts of the reports

FIGURE 4.3. Dr. Reina Torres de Araúz with Guna, Panamanian, and U.S. representatives preparing to meet to execute payment for trees and vegetation removed during the construction of a weather-recording station in conjunction with the canal studies in the Darién, Soskatupu, Comaro de San Blas, September 20, 1966. APICSC Scrapbook, RG 220, U.S. National Archives and Records Administration, College Park, Md.

addressing the sea-level canal's foreign policy, national defense, and financial dimensions, among other topics.[81]

Problems persisted nonetheless. Despite carefully staged ceremonies, protests against both Yankee imperialism and the oligarchic Robles government erupted on the third anniversary of the Flag Riots, January 9, 1967.[82] Two months later, the leaking of the almost identical texts of the Panama and Colombia site survey agreements renewed angry rumors that the Route 25 proposal was meant only to weaken Panama's hand in the treaty talks.[83] At a subsequent congressional hearing, the corps representative Woodbury testified that "agitators" were sowing discontent among the Indigenous people working for the Route 17 survey by telling them that they were not being paid enough. Woodbury insisted that the rate, 37.5 cents per hour, was normal and that higher salaries might disrupt the host country's economy; he also praised the Panamanian government for its "great help in keeping these difficulties under control."[84] The Office of Interoceanic Canal Studies struggled to counter the unfavorable publicity, even as it was forced by insufficient funds to begin phasing out the Route 17 studies prior to the start of the 1968 dry season.[85]

More canal study calamities were unfolding stateside. For the second year in a row, President Johnson acceded to pressure from the State Department and the Arms Control and Disarmament Agency to call off the Cabriolet test. After canceling it in early 1966, Johnson agreed to another delay in February 1967 to avoid disrupting negotiations in Mexico City among twenty-one countries to outlaw nuclear weapons in Latin America via the Treaty of Tlatelolco. It was a tough call because the AEC had already announced the test, which led a member of the congressional committee on atomic energy to accuse the president of having "caved into pressure from a noisy group of liberals who urge us to go to any extreme to obtain disarmament treaties."[86]

However, the administration could not afford the risk of releasing radioactivity into Mexico. Underground explosions at the Nevada Test Site from 1963 to 1966 had emitted fallout eight times. Even though none of the incidents involved cross-border releases, the State Department had just sent a representative to the Tlatelolco treaty conference to protest a provision that allowed participating nations to use nuclear devices for peaceful purposes.[87]

To placate his pro-PNE critics, Johnson asked Congress to increase the Plowshare appropriation from $15.7 million to $19.5 million for the fiscal year beginning on July 1, 1967. But the program was attracting more and more adverse publicity. Prominent defense and scientific advisors went on the record to denounce "the so-called Project Plowshare" as an endeavor that might promote weapons development as it reduced public works construction costs, an unjustifiable trade-off.[88]

Anderson's two-and-a-half years of "poker diplomacy" culminated in another demoralizing setback for the commission.[89] In June 1967, Presidents Johnson and Robles announced that the two nations had agreed to replace the Hay–Bunau-Varilla pact with three new treaties governing the existing canal, the Zone defense bases, and the proposed sea-level canal. But before the accords could be signed, the *Chicago Tribune* published the unofficial texts. Political firestorms erupted in both countries: conservatives in the U.S. responded to the "surrender in Panama" with outrage, as did Panamanian students, for whom the treaties did not go far enough toward ensuring their country's economic and political independence.[90] Anderson and other administration members held out hope that the diplomatic process could continue.[91] But Representative Flood unleashed a new wave of rebukes, threatening to lead 150 representatives to the Senate to disrupt any ratification hearings. "The moving line of indignant Congressmen," pronounced the *Wall Street Journal*, "would be just one more strand in a web of problems besetting one of the world's most ambitious engineering projects. . . . The web of obstacles—political, diplomatic, scientific—may in the end make the new canal more difficult to build than the lock canal completed in Panama 53 years ago."[92]

Conclusion

Plowshare proponents took all the setbacks in stride, remaining confident that if only politics and emotions could be compartmentalized, PNEs would take their rightful place as the world's construction method of choice.[93] An August 1967 report on the economics of peaceful nuclear excavation by a data analytics firm for the Atomic Energy Commission asserted that history was on the agency's side: "Regarding the general decision whether or not to use Plowshare at all, a point worth mentioning is of historical nature: *no new technology that has ever become available to man has been rejected*."[94]

A deeper dive by the contractors into the history of technology would have provided ample evidence to the contrary. After chemical weapons wreaked havoc in World War I, for instance, most of the world's nations outlawed them via the Geneva Protocol of 1925. (The United States did not ratify it until 1975, but Kennedy and Johnson officials insisted their use of herbicides in Vietnam did not violate the protocol because the chemicals killed plants rather than people.[95]) Likewise but less spectacularly, electric vehicles and solar technologies had enjoyed only brief stints up to that point of the twentieth century.[96] In all these cases, governmental decision makers—often working with elite economic stakeholders—played powerful roles in determining which new technologies and associated infrastructural systems gained dominance.

For those still hoping in the late 1960s that PNEs would become a routine tool for harbor and canal construction, support came from an unexpected source. The cautionary *Bulletin of the Atomic Scientists*, the periodical that since 1947 had featured the famous Doomsday Clock to represent the changing threat to humanity by nuclear technology, published a special report in its December 1967 issue. While "it is not necessary to use nuclear explosives to construct a sea-level canal," wrote the authors, the technique appeared feasible: "The problems of blast-damage and radioactivity are inconvenient, and they limit the choice of a route to remote areas, but these are manageable problems."[97]

Unbeknownst to the *Bulletin* authors, however, the CSC was wrestling with many problems and inconvenient truths. Managing public relations and governmental expectations, and reconciling their faith in the potential of nuclear

At the same time that they quietly grappled with the clay shales problem, the commissioners faced intensifying calls to expand the scope of the bioenvironmental feasibility studies. The CSC had always accepted the need for biological data to predict the effects of nuclear excavation on the human societies of the isthmus. But they drew a sharp line between essential and desirable bioenvironmental information when the zoologically oriented canal treaty negotiator, John Irwin, suggested looking into the broader effects of a sea-level waterway on oceanic organisms. By the late 1960s, the Anderson Commission could no longer ignore the marine biologists pressing for a share of the research funds earmarked for what Ira Rubinoff, the author of the 1965 article that had likely sparked Irwin's interest, now deemed "the greatest biological experiment in man's history."[6]

Almost a decade earlier in Alaska, the AEC officials promoting the Project Chariot nuclear harbor had tried to co-opt local biologists by providing research contracts.[7] But the Anderson Commission could barely afford the studies it considered crucial, let alone support esoteric inquiries about marine faunal exchange. Besides, had not the Panama Canal joined the Atlantic and Pacific decades earlier without unleashing waves of disruptive biological invasions? Had not sailors been transporting marine life inside and on their ship hulls for *centuries*?

Having already invested three years and most of its budget in an immense set of studies pertaining to shipping, foreign relations, national defense, and nuclear engineering feasibility, the commissioners were in no mood to bankroll what museum-and field-based naturalists of the Smithsonian Institution and other marine biology programs sought: a ten-year baseline study to elucidate the evolutionary and ecological effects of breaking a land barrier that had separated two oceans for some three million years. How they handled the pressure highlights one of the most intriguing side effects of the sea-level canal story, the emergence of an international forum for debating the effects of maritime transportation on marine biological diversity.

The Ivory Soap Bomb and the Isthmus

After two years of waiting for the canal cratering experimental program to begin, the Anderson Commission received the wonderful news in January 1968 that President Johnson had finally permitted the 2.3-kiloton Cabriolet shot to proceed at the Nevada Test Site. Because radioactive fallout might escape into the atmosphere and contaminate the local milk supply or cross the border to cause an international incident, the timing had to be precise—after the State of the Union address, but before the Nevada grazing season. In a stroke of good luck for the advocates of Plowshare, the winds on the day of the test blew away from Mexico, and a snowstorm prevented debris from reaching Canada in detectable amounts.[8]

Although the Cabriolet test created a crater of only 360 feet wide and 120 feet deep, a follow-up experiment in March generated exciting results for proponents of the nuclear waterway. Project Buggy, a simultaneous detonation of five nuclear explosives (spaced 150 feet apart at a depth of 150 feet), produced what Representative Chet Holifield of the Joint Committee on Atomic Energy called a "miniature-size canal in the Nevada desert" (300 feet wide by 80 feet deep by 900 feet long). His assessment of the situation as "very hopeful, very promising" renewed calls to amend the Limited Nuclear Test Ban Treaty "to reconcile it with common sense."[9] More favorable publicity appeared in news reports that fallout from both shots occurred only a few hundred yards from the craters, and that no radiation could be detected three days afterward, unlike earlier tests. An unnamed AEC source attributed the smaller amount of radiation to the "really tiny fission device" used to trigger the thermonuclear explosive. For the "friends of Plowshare," the 97 percent fallout-free device—dubbed the Ivory Soap Bomb—offered great reassurance: "We've been talking about doing it for years, but now we could really get down and dig a second Panama Canal with atomic explosives."[10] The long political delays had served their purpose of giving the science and technology of PNEs time to catch up with expectations.

Or so it seemed. Such rhetoric discounted the continuing opposition by Latin Americans to the use of nuclear explosives on their lands. Panamanian ambassador Jorge T. Velasquez, who called instead for expanding the existing waterway, announced, "We know that experiments have not shown there is control over the dangers of atmospheric and underground radioactivity, nor the practical possibility of this technique in the construction of a sea-level canal."[11] Not only were the AEC's assurances about clean PNEs wearing thin, so was the techno-economic rationale for building a second waterway. The U.S. Navy's largest ships needed to be able to cross the isthmus as quickly as possible, but that was not the case for an emerging class of commercial vessels. In 1966, massive new tankers, some capable of carrying 2.2 million barrels of crude oil from the Middle East to Europe, had already begun bypassing the Suez Canal to save on tolls; voyaging around the Cape of Good Hope took longer, but the size of their cargoes made up for the loss of time.[12] That same year, the burgeoning Japanese shipbuilding industry completed a 215,000-ton tanker capable of drawing fifty-six feet, which exceeded the Panama Canal's maximum channel depth of forty feet, and analysts predicted that tankers of 500,000 and even 1 million tons would soon be feasible.[13] The global shipping industry was changing in ways that demonstrated the need for the Panama Canal to adapt or else become a mere monument to a bygone era—but that also undercut the long-standing arguments for a sea-level channel.

As the Anderson Commission's economic advisors tried to predict future shipping trends, the Route 17 geological consultants delivered unwelcome news. Drilling samples verified that about half of the route, a twenty-mile stretch in the Chucunaque valley, crossed land whose bedrock consisted of clay shales (rather than basalt, the much harder material underlying most of the Continental Divide, the assumption of which had underpinned the 1964 selection of routes). Nuclear explosives would likely not form stable slopes in such an unstable medium, meaning that achieving the shallow slopes needed to prevent the crater from collapsing over time would require conventional excavation techniques.[14]

Clay shale happened to be the same type of sedimentary rock that had caused destructive landslides during the construction of the original Panama Canal, and thus the problem was not a surprise to engineers familiar with the physics of the isthmus. As one of the leaders of the Army Corps of Engineers' Nuclear Cratering Group explained at the March 1968 commission meeting, "The original thinking was that if we ran into a bad actor like clay shale along the routes, we could approach it from a conventional excavation standpoint."[15] However, the prospect of bulldozing such a large area undermined the financial advantage PNEs had offered in the first place.

While engineers consulted with leading soil mechanics specialists to determine whether the geological discovery really did rule out PNEs, the Livermore physicists conducted chemical explosive experiments to try to model the effects of burying thermonuclear devices in wet clay shale. John Kelly of the AEC's Division of Peaceful Nuclear Explosives later even discussed the matter with his Soviet counterparts, who insisted it was not a problem.[16] A few days after the March 1968 CSC meeting at which the issue was first unveiled, commission member Milton Eisenhower vented his frustration in a private letter to Executive Secretary Sheffey that the group's most vociferous member, Raymond Hill, already appeared ready to abandon Route 17: "When we began our deliberations, we were enthusiastic about atomic construction. Now we are on the verge of deciding that, largely because of possible conditions of soil instability, atomic excavation is not practical . . . I don't want to be drawn inexorably to route 10 and a sea-level canal by traditional construction methods . . . merely because his expert opinion points in that direction."[17] Route 10 referred to a new potential seaway site about ten miles west of the existing canal, to which the commission had begun devoting attention in the fall of 1967 as an alternative to the nonnuclear option of converting the existing canal.[18]

If nuclear excavation of Route 17 was no longer possible, then that had major implications for future U.S.-Panama treaty negotiations. The 1967 draft treaties had established the U.S. commitment to building the sea-level canal in Panama and had granted Panama the right to veto nuclear methods. However, as outsider analysts explained, if the existing waterway continued its descent into obsolescence, and if PNEs constituted the only financially feasible means of constructing a new passage, then Panama would pretty much have to agree to a U.S.-built nuclear canal.[19] To preserve the U.S. upper hand in future negotiations, how much should the commission reveal about the clay shale problem?

The deadline for the fourth annual report, covering the period from July 1, 1967 to June 30, 1968, was approaching, and it would be the last one prepared for President Johnson, who had decided not to run for reelection. The previous three reports had already erred on the side of vagueness, and thus to mollify the commission's critics while preserving as much U.S. leverage as possible, Sheffey recommended a selective, semitransparent approach: it should "start surfacing some of the problems we foresee and not catch the Congress by surprise when the final report comes in. The problem is that we don't want to disclose things that will handicap treaty negotiations in the future by our action."[20] But reconciling such opposing goals proved impossible. In conveying the ensuing report to Congress, President Johnson omitted the clay shales matter from his public statement, and sent mixed messages by emphasizing both the benefits of the conventional Route 10 and the favorable results of the two 1968 Nevada Test Site shots.[21] He also overlooked another unwelcome issue consuming more and more of the commission's attention: regardless of the method used to cut the channel, what would happen once the flora and fauna of the two oceans reunited after a few million years of evolutionary separation and speciation?

Marine and Evolutionary Biologists Barge In

The secretary of the Smithsonian Institution, Leonard Carmichael, had posed a version of this question to AEC director Glenn Seaborg back in the spring of 1963, when U.S.-Panama tensions, the resumption of nuclear testing, and the negotiations over the test ban treaty generated widespread publicity for the Panatomic Canal idea.[22] Carmichael had likely been prompted by an employee of the Museum of Natural History, oceanographer I. Eugene Wallen, who had spent years studying the faunal effects of nuclear testing in the South Pacific.

FIGURE 5.1. U.S. president Lyndon B. Johnson presenting a plaque to Smithsonian Institution secretary S. Dillon Ripley, June 13, 1967. Ripley and other Smithsonian officials tried but failed to secure federal funds from Johnson's canal study commission for a ten-year baseline inventory of marine life on either side of the proposed seaway. Smithsonian Institution Archives, 92-1656.

Carmichael wound up dropping the offer to help conduct baseline taxonomic and ecological research along the proposed canal routes, but his successor, ornithologist S. Dillon Ripley, saw a great opportunity (fig. 5.1).[23] Determined to improve the Smithsonian's waning reputation for biological excellence, Ripley reestablished contact with the AEC in the summer of 1964.[24] Later that year, days after President Johnson announced the sea-level canal plans, Wallen submitted a $2 million proposal "to determine the potential damage by canal construction to the populations, distributions and abundances of marine and terrestrial organisms on the two proposed routes" in Panama and Colombia.[25] However, the AEC chose the Battelle Memorial Institute, a private organization that in turn subcontracted with university-based teams.[26] Deeply disappointed, Wallen and Ripley regrouped to consider their options.[27] One of the veterans of the Chariot affair, AEC Environmental Sciences Branch chief and plant ecologist John Wolfe, provided an encouraging boost: "In such massive engineering proposals," he wrote Wallen in reference to the sea-level canal, "biology is no longer a flower girl, she's the bride."[28]

For Ripley, the canal feasibility studies held the promise of attracting new funds not only for the Museum of Natural History, but also for one of the Smithsonian's far-flung research facilities. Decades earlier, in 1937, he had visited Barro Colorado Island (BCI) in Panama's Canal Zone while voyaging from Philadelphia to New Guinea for a zoological expedition.[29] BCI was an artifact of the construction of the canal, as the reservoir created by Gatun Dam left only the highest hilltops of the Chagres River valley unsubmerged. The island was also a testament to the Smithsonian's embrace of the isthmus as a model study site, beginning with the prescient 1910–12 biological survey of the Panama Canal Zone, which generated several studies documenting the area's marine and terrestrial biota.[30] The survey led the governor of the Canal Zone to designate BCI a reserve in 1923, and over the next two decades, the research station drew more scientific visitors than any other tropical research facility.[31] BCI and the associated Canal Zone Biological Area (CZBA) became a Smithsonian bureau after the war and by the late 1950s had provided the setting for hundreds of articles on tropical forest biota and dynamics. However, the CZBA lacked facilities for long-term terrestrial and marine research. Expanding the institutional base of tropical biology gained urgency as tropical nations decolonized and sought to accelerate deforestation for development.[32]

Another influential source of encouragement for expanding the CZBA was Ernst Mayr, an old friend of Ripley and fellow ornithologist, the head of Harvard's Museum of Comparative Zoology, and the mentor of BCI director Martin Moynihan. Mayr had gained fame in the 1940s as a proponent of the Modern Synthesis of evolutionary biology and the concept of allopatric (geographic) speciation, a topic for which the Panamanian isthmus provided an ideal research setting.[33] Knowing the approximate period of the geological emergence of the Central American land bridge made it possible to date the origin of evolutionary differences between marine species on either side of the isthmus.[34] The 1910–12 Smithsonian survey had built upon a few pioneering studies to expand knowledge of the marine fishes of the Atlantic and Pacific coasts and associated evolutionary effects of geographic isolation.[35] Naturalists subsequently documented some other taxonomic groups west and east of the isthmus that appeared similar but were not identical, as Mayr substantiated for shallow-water sea urchins in a foundational paper that called for more research on modes of speciation in marine organisms. But like several other such researchers, he based his results not on a visit to the biogeographical barrier in question, but rather on an earlier taxonomist's work.[36]

Mayr's recognition of the evolutionary importance of the Central American isthmus shaped his mixed reaction to President Johnson's December 1964 announcement. As he expressed to a colleague on the President's Science Advisory Committee, the consequences of conjoining two distinct oceans raised daunting questions beyond the issue of radioactivity:

> I have been worrying for some time about the contemplated sea-level canal across the Isthmus of Panama. There is little doubt that such a salt water connection between the two great oceans will have many and drastic effects on the marine faunas and floras. There are closely related species living on either side of the Isthmus and no one can predict with certainty what will happen if such species come in contact with each other. Will one wipe out the other, or will it hybridize with it? How many disease organisms will be carried from one ocean into the other? What will the tidal currents do in the canal? Will the inflow of the more silty waters of the Pacific damage coral reefs on the Atlantic side? I do not know what the situation is in commercial fisheries, and other utilisation of marine organisms, but . . . I am sure that the opening of the canal will produce many problems, and world science would never forgive us for not being prepared for such eventualities.[37]

In addition to expanding marine stations in Panama, Mayr argued the United States should appoint a board to oversee the research needed to answer such questions, perhaps involving the National Academy of Sciences. In this regard, he noted, "I might add parenthetically that organismic biology is, on the whole, poorly represented on high level boards in Washington. This is not injurious in most instances, but it may lead occasionally to the neglect of an important problem, as I believe is the case with respect to the Panama Canal."[38] Mayr's aside was more than parenthetical, for he had been promoting organismic biology as a means of counteracting the growing dominance of molecular approaches to the life sciences.[39] Also, conservation was not a prominent theme of his career, though he had published a few articles decades earlier about bird protection.[40] Mayr's interest in both the ecological risks and research opportunities of the proposed waterway spoke to the spectrum of positions among the biologists who would play major roles in the sea-level canal story.

Mayr had an important informant who helped shape his thoughts about the sea-level canal—his recent graduate student Ira Rubinoff. During visits to the CZBA in 1961 and 1962, Rubinoff had begun collecting and breeding fish from the Caribbean and Pacific coasts to delineate evolutionary divergence in species separated by the isthmus. Conducting such research in the Canal Zone entailed unusual occupational hazards, from having to pass checkpoints manned by surly sentries to missing optimal tides due to the lack of gas stations that serviced military vehicles on weekends.[41] But the Canal Zone provided protection and amenities for North American and European researchers, and the efforts to expand the CZBA paid off in 1965, when Moynihan secured leases with the U.S. military for two Atlantic and Pacific marine stations, the Smithsonian hired Rubinoff as assistant director for marine biology, and Congress raised the appropriation from the original $10,000 to $350,000.[42] Ripley announced the name change to Smithsonian Tropical Research Institute (STRI) the following year.[43] As STRI's marine director, Rubinoff provided an energetic on-the-ground complement to the D.C.-based Smithsonian campaign of Ripley, Wallen, and Assistant Secretary for Science Sidney Galler, a former Office of Naval Research biologist.

In July 1965, Rubinoff published the first article on the sea-level canal's potential nonradiological effects, the *Natural History* essay that likely caught Irwin's attention.[44] Several outcomes seemed possible, including large-scale extinctions, as suggested by principles of ecology and genetics, morphological comparisons of Caribbean and Pacific shore fishes, and historical human-facilitated introductions of organisms to new areas. Dutch elm disease and the Australian rabbit invasion were infamous in the annals of pest outbreaks; other cases like the canal-facilitated migration of voracious Atlantic sea lampreys into the U.S. Great Lakes were less well known but equally devastating to local ecosystems and fishing-based communities. Rubinoff assured readers that the sea-level canal would "not provide every species with a free pass to a new ocean." But why leave anything to chance? A strategic research program such as that of the International Indian Ocean Expedition, one of several large-scale geophysical initiatives of the Cold War era, would begin to demystify the uncertain effects of marine faunal mixing.[45]

Risky Mix

Scientific conferences provided an important venue for addressing the sea-level canal's ecological and evolutionary consequences. At the 1965 International Conference on Tropical Oceanography in Miami, Rubinoff and oceanographers from other distinguished institutions discussed the importance of expanding the preproject studies beyond radioactivity in the food chain, and marveled at what they considered the myopia of attendees from the Battelle Memorial Institute and other AEC subcontractors.[46] However, in the spring of 1966, Battelle Memorial Institute officials invited Galler to their Ohio headquarters to discuss the "myriad problems" that had become apparent as they realized the limitations they faced in fulfilling their AEC obligations.[47] Galler convinced the Battelle managers to invite the Smithsonian to cosponsor an international symposium aimed at developing a long-term ecological survey for the Canal Study Commission.[48]

Planning the symposium required a great deal of strategizing.[49] In August 1966, Galler convened a meeting at the Museum of Natural History of all the relevant organizations: Battelle (represented by radioecologist William Martin), the AEC's Division of Peaceful Nuclear Explosives (John Kelly) and Environmental Sciences Branch (John Wolfe), and the Army Corps of Engineers and CSC (jointly represented by Harry Woodbury).[50] The group, which also included representatives of the American Institute of Biological Sciences and the Organization of American States, discussed the task force's draft proposal for a $6 million five-year precanal marine survey, and agreed that holding an international symposium of tropical biologists in Panama would help identify and publicize the top canal-oriented research priorities.

However, sharp differences emerged regarding questions of framing and language. Pointing to a section of the draft that drew on Rubinoff's hybridization studies of Atlantic and Pacific fish, Woodbury noted that agricultural scientists "devote their lives to developing hybrids." Moreover, because some of the transisthmian species liable to go extinct might be "those we would want to exterminate," attention should also be devoted to the beneficial environmental changes that the canal might effect. When a Smithsonian employee explained that only by accentuating the negative would Congress respond with the needed funds, Woodbury stated that using words like "catastrophe" and "profound disturbance" would upset his bosses, who had been tasked with determining where—not whether—to build the sea-level canal.[51] Colonel Sheffey of the CSC likewise argued in a follow-up letter that the task force dwelled too much on "unfavorable ecology changes."[52] Galler tried to smooth things over by saying it was just a draft, but the exchange revealed the fine line between courting and alienating the oversight agencies.

The Smithsonian team also recognized the importance of "not antagonizing the Panamanians," and thus they invited both U.S. and Panamanian representatives to the symposium slated for November 1966. They also emphasized the goal of advancing basic ecological research.[53] Yet the sea-level canal featured prominently in the discussions of the four-day symposium (attended by sixty North American and Latin American biologists) and in the ensuing write-up for *BioScience*.[54]

Allocating internal resources to stage the Panama conference and relevant pilot projects at STRI became an important part of the Ripley team's evolving strategy to create, in Galler's words, "a receptive constituency" among the scientific and governmental communities to fund the Smithsonian's long-term program.[55] Smithsonian research awards enabled the husband-and-wife team of Ira and Roberta Rubinoff to focus on interoceanic colonization and isolating mechanisms in Atlantic and Pacific fish populations.[56] Such grants also made it possible for visiting marine biologists to conduct research at STRI's new facilities on the Caribbean and Panama Bay coasts, resulting in publications that would drive much of the sea-level canal debate.[57]

The Smithsonian's self-funded canal studies also benefited marine scientists on the other side of the globe. In early 1967, STRI joined the Hebrew University of Jerusalem to investigate Red Sea and eastern Mediterranean biota. The hundred-mile-long Suez Canal, completed in 1869, had a high-salinity interior that blocked most, but not all, species' migration between the Red Sea and Mediterranean.[58] The Smithsonian–Hebrew University initiative made the most of a new U.S. program to apply excess foreign currencies generated by the sale of agricultural surpluses toward Smithsonian research in excess-currency countries, and Israeli scientists gained full access to the Suez in June 1967 following the Six-Day War.[59] Israeli marine biologists made major contributions in the ensuing decade to the study of Red-Med species exchange, cleverly dubbed "Lessepsian migration" after Ferdinand de Lesseps, the architect of Suez and the failed 1880s-era Panama Canal.[60] Biogeographic studies of the Suez Canal provided an important and disputed point of comparison and contrast for researchers endeavoring to predict the ecological and evolutionary effects of the Central American sea-level waterway.

The Panama conference seemed to be a major stepping stone, as exemplified by the warm letters exchanged afterward by Ripley and Woodbury, who had jokingly referred to himself in his presentation as a "devastation engineer."[61] Throughout 1967, Smithsonian and CSC representatives corresponded and met on occasion in Washington to discuss how to acquire $10 million for a decade-long Smithsonian-directed ecological program.[62] Yet despite numerous presentations on the topic to House and Senate appropriations subcommittees, Galler complained, "Congress thinks of the SI primarily as a museum of public exhibits."[63] Even so, by early 1968, the tide seemed about to turn, as Smithsonian personnel prepared new proposals for consideration by several public and private organizations.[64]

The Smithsonian's target constituency for canal ecology fundraising did not include the public, but mainstream attention to the nonnuclear dimensions of the sea-level canal was increasing at the time Galler made his confident prediction. At the December 1967 meeting of the American Association for the Advancement of Science in New York City, LaMont Cole, the president of the Ecological Society of America, asked the shocking question, "Can the world be saved?" His speech invoked the sea-level canal as one of several examples of human technological folly, and appeared in both *BioScience* and the *New York Times Magazine*. For the Smithsonian scientists, however, his focus on radioactivity and his erroneous claim that the Pacific Ocean stood higher than the Atlantic by an average of six feet, evinced the importance of advancing a more informed debate.[65]

To that end, in August 1968, Rubinoff published an article in *Science* with the provocative lede, "An opportunity for the greatest biological experiment in man's history may not be exploited."[66] The article expanded on the earlier *Natural History* piece by quoting from the English ecologist Charles Elton's 1958 book *The Ecology of Invasions by Animals and Plants* to discuss "ecological explosions" of noxious, nonnative species. New York's Erie Canal, and later Ontario's Welland Canal, had enabled predatory Atlantic sea lampreys to spread throughout the Great Lakes, decimating the commercial fisheries.[67] Yet even such extensive losses, along with those inflicted in the United States by Japanese beetles, gypsy moths, fire ants, and chestnut blight, would likely be dwarfed by the sea-level canal's unleashing of "mutual invasions of Atlantic and Pacific organisms [which] should be much more extensive, numerous, and rapid."[68]

The article drew not only on historical case studies but also on the Rubinoff laboratory's experimental research. To test the possible consequences of interbreeding between formerly isolated populations, Rubinoff and his collaborators, including his spouse Roberta Rubinoff, conducted mating experiments between Atlantic and Pacific marine gobies. The range of results regarding the viability of the offspring, and the broader question of which species might dominate, hybridize, or go extinct, demonstrated the need for extensive knowledge of the morphological and behavioral characteristics of the organisms in question, as well as their degree of genetic divergence over time. Moreover, according to well-established ecological principles regarding competition between species for the same resources, the struggle for existence would inevitably drive some organisms to extinction; the changed physical conditions induced by the sea-level canal would also disrupt long-standing population dynamics. Because the seaway would extinguish many species that might not ever have been collected and described by scientists, a preproject inventory would provide "a potentially important historical base for biological oceanography of the future."[69]

Rubinoff closed the article by quoting from a 1963 speech that President John F. Kennedy had delivered to the National Academy of Sciences in celebration of its one hundredth anniversary. Two weeks after signing the Limited Nuclear Test Ban Treaty, Kennedy used the academy event to announce his intention to minimize government secrecy and thereby foster greater discussion by the international scientific community about proposed large-scale development projects. Furthermore, he warned that while "the problem of conservation" had previously entailed the destructive use of natural resources, "science today has the power for the first time in history to undertake experiments with premeditation which can irreversibly alter our biological and physical environments on a global scale."[70] Along these lines, Rubinoff called for the government to provide the support needed to quantify and predict the biological effects of the proposed sea-level canal, and to establish an independent, multidisciplinary "control commission for environmental manipulation" to regulate all proposed megaprojects.[71] In important ways, he echoed the drafters of what would become the U.S. National Environmental Policy Act of 1969, the groundbreaking law requiring preliminary environmental impact assessments of federally funded projects.[72]

Rubinoff tried to leverage what would now be called a "traction opportunity" to advance both a specific research program and a more environmentally sensitive approach to development and policy-making.[73] However, all the organizations that he and Smithsonian leadership approached, including the Ford Foundation, National Institutes of Health, Office of Naval Research, Air Force Office of Scientific Research, and Naval Oceanographic Office, turned down the proposal for a decade-long precanal investigation.[74] Many complex reasons accounted for the unwillingness of public and private organizations to commit to such an ambitious research endeavor. Throughout the twentieth century, the institutional status of naturalist-based disciplines such as taxonomy, biogeography, and evolutionary and ecological science lagged further and further behind the more reductionist, experimental life sciences, as epitomized by molecular biology. Moreover, grant programs for ecology tended to focus on short-term, small-scale studies in places far from human influence.[75]

Rubinoff's work also challenged persistent attitudes about marine resilience and invulnerability to human-mediated change. Not until the 1990s and early 2000s did funding agencies begin devoting serious attention to ocean conservation science. A terrestrial bias also permeated the agenda of the emerging environmentalist movement. Fears of marine species migration and mixing did not fit the agenda of potential allies who focused on the terrestrial threats posed by synthetic chemicals, rapid population growth, overconsumption, and other toxic byproducts of technological modernity. And even though oil spills and overhunting presented obvious dangers to fish and marine mammals, competition from nonnative introduced species was much harder to comprehend, for both mainstream scientists and environmental advocates.[76]

Indeed, the late 1960s was still an era of intentional aquatic species introductions. For over a century, governments and acclimatization societies in search of cheap protein or recreational opportunities had shipped live fish and shellfish for transplanting in rivers and bays around the world. Pests often moved with the desired species, resulting in occasional warnings, most famously from Elton, who described oysters as the "sessile sheep" of the sea.[77] Nevertheless, government officials stocked Lake Michigan with Pacific salmon in 1966 and replaced France's adopted oyster, the Portuguese *Crassostrea angulata*, with the Japanese *C. gigas* from 1971 to 1975.[78] Although a reckoning with the ecological costs of such actions was dawning on biologists, questioning the wisdom of transplanting aquatic species still flew in the face of a hundred years of scientific advice.[79]

Another difficult thing to explain pertained to the existing Panama Canal's failure to facilitate disruptive marine species exchange. As Elton had observed in the sole marine chapter of *The Ecology of Invasions by Animals and Plants*, "accidental carriage in or on shipping, that is in water ballast tanks or on the hull, has been a powerful and steady agency dispersing marine plants and animals about the world."[80] Yet thousands of ships had transited the Panama passage since 1914, and only one ocean species, the euryhaline Atlantic tarpon, had ever appeared capable of breaching the fresh waters of Gatun Lake.[81] Of course, identifying invasive species required consistent monitoring as well as thorough knowledge of the native organisms of each coast. The Rubinoffs had recently discovered breeding populations of an Atlantic goby in the brackish Miraflores Third Lock Lake near the Pacific side, which suggested that further surveying might identify additional "successful amphi-American migrations."[82]

The CSC had thus far maintained cordial relations with the Smithsonian in the joint initiative to expand federal funding of nonradiological research, but in the fall of 1968, Executive Director Sheffey penned a letter to the editor of *Science* taking issue with Rubinoff's article. The commission was already funding studies to delineate the seaway's possible biological consequences, and besides, marine organisms could already transit the Panama Canal via three methods: by swimming and drifting through the locks and freshwater interior, by clinging to the sides of ship hulls, or by being swept up into the ballast tanks of ships on one side and then released on the other. Biotic exchange must thus already be occurring on a large scale, without any ill effects. While a sea-level waterway might permit larger swimming and drifting biota to pass through, asserted Sheffey, "the area of danger of harmful biological changes when the oceans are joined is much less broad than it first appears."[83]

Sheffey's attitude reflected the conventional wisdom of the shipping industry, that maritime transportation had already mixed the world's marine biotas and that any resulting biological harm was only localized—and more than offset by the enormous economic benefits of global shipping. Rubinoff responded to Sheffey's letter that too little evidence existed to support his claims, and that he sought not to cause alarm but rather "to provoke action by the scientific community to predict the probable results of permitting two separate biotas to merge, and to measure such changes as may occur when and if the canal is dug."[84] Such careful language reflected the difficulty of reconciling the worldviews of hydroengineers and scientists, and of meeting traditional expectations of scientific objectivity amid the environmental policy-making upheavals of the 1960s. The concept of "a risky mix" applied not only to the environmental threats of intermixing the fauna and planktonic flora of two oceans.[85] It also involved the professional hazards facing scientists who stepped too far outside their idealized realm of social isolation.

Biological Disaster or Grand Evolutionary Experiment?

Rubinoff's 1968 article opened the floodgates of critique from fellow biologists. The major intellectual disputes encompassed conflicting interpretations of the limited empirical research on transisthmian evolution, biological diversity, and biotic mixing, and of the validity of the Suez and Panama Canals as models for predicting how the sea-level waterway might affect ecological and evolutionary processes. Some participants also questioned Rubinoff's more explicitly political assumptions and framing choices. At a time when the negative environmental consequences of modern technoscience were gaining more and more attention, but when marine biodiversity and bioinvasions barely registered as issues of concern, the debates reveal how biologists of different disciplinary backgrounds and levels of conservationist concern struggled to communicate the political relevance and fund-worthiness of a topic for which they lacked a common vocabulary and conceptual understanding.

The first published scientific response to Rubinoff's essay came from the ichthyologist and marine zoogeographer John C. Briggs.[86] While appreciating the redirection of scientific attention away from radiation damage, Briggs took issue with Rubinoff's sense of opportunism and inevitability. The breaching of the isthmian zoogeographic barrier would, he predicted, allow animals of the apparently more species-rich western Atlantic ecosystem to dominate their relatives in the eastern Pacific, eventually wreaking havoc on the latter: "Let us *not* be concerned about preparation for a great biological experiment. The important question is: Should the sea-level canal project be undertaken at all? Are we prepared to assume the responsibility for the irrevocable destruction of several thousand unique species in the Eastern Pacific?" For Briggs, the sea-level canal proposal posed "a conservation problem of an entirely new order of magnitude."[87]

The absurdity of conducting routine baseline research in the face of an extinction crisis led Briggs to publish a much longer article in the January 1969 issue of *BioScience* with the striking subtitle "Potential Biological Catastrophe." He argued that the Suez offered disturbing empirical evidence of how organisms from a richer marine ecosystem might outcompete the natives of a less stable one, as shown by the aggressive migration of Red Sea fishes, crustaceans, mollusks, and tunicates into the empty niches of the nutrient-poor eastern Mediterranean. Briggs calculated that 6,720 Caribbean species would migrate westward through a Central America sea-level canal and 4,480 less robust Pacific species would move eastward, resulting in massive extinctions among the latter. While suggesting that a sea-level waterway could be designed to kill migrating animals with chemicals or hot water, Briggs deemed such methods "risky and distasteful" and instead called for improving the existing canal to meet the needs of world shipping.[88]

Briggs's analysis provoked strong reactions from fellow members of the scientific community. Rubinoff and other scientists questioned his statistical methods and assumptions.[89] Biological oceanographer Gilbert Voss later implicitly called Briggs and Mayr "two of the most outspoken alarmists [who] have had no personal experience in the areas and have not engaged in research relating to the problem."[90] And the eminent ichthyologist Carl Hubbs, whose research on the native and nonnative fishes of both the Great Lakes and Suez Canal dated to the 1920s, probably had Briggs in mind when he urged the president of the National Academy of Sciences in February 1969 to appoint a committee to address the questions raised by the sea-level canal. In contrast to "the irresponsible nature of some of the published discussions," Hubbs emphasized, "I am deeply impressed with the potential biological consequences, for better or for worse, and with the fantastic opportunities for research that are presented," points he reiterated at oceanographic and zoological meetings in Curaçao and Caracas.[91] Hubbs's points underscored the failure of biological oceanography and marine biology to keep pace with the advances in physical oceanography made possible by Cold War military patronage of the earth sciences.[92]

Hubbs also drew on emerging research by the Smithsonian-Israeli collaboration to support his assessment of the Suez as "an imperfect model" for estimating the biotic effects of the proposed Central American canal.[93] The two isthmuses differed in terms of not only salinity levels but also geological histories. The more recently separated coastal Caribbean and Pacific biotas contained many closely related pairs of species, whereas the Mediterranean and Red Sea biotas contained almost no such sister species, differences which presented "extremely urgent, and at the same time very promising" opportunities for systematic, ecological, and evolutionary studies.[94]

A stark rejection of the Suez situation as a valid precedent came from a STRI predoctoral visitor, Robert Topp. It was no surprise, he argued, that Red Sea ichthyofauna had occupied the vacant niches of the impoverished Mediterranean. By contrast, because most of the ecological niches on either side of the Central American isthmus were already filled, species introgression through a sea-level canal would likely not cause widespread extinction. "Faunal enrichment" might even occur, especially in the Caribbean.[95]

In support of his prediction that the proposed waterway would not cause widespread extinctions, Topp mobilized new evidence made possible by other visiting researchers to Panama and the Smithsonian's internal funding. After towing cheesecloth-wrapped marine animals through the canal, Robert Menzies concluded that genetic exchange was probably already occurring on a large scale due to the ability of fouling organisms attached to the bottoms of ships to survive the full fifty-mile transit.[96] Another visiting biologist studying marine plankton transport, Richard Chesher, confirmed via interviews with Panama Canal Company officials that tankers and freighters had been required since 1956 to ballast down prior to the transit to ensure maneuverability.[97] "Much biotic transfer" was thus probably already occurring as ships made the eight-hour journey and then emptied their ballast tanks on the other side, to no ill effect.[98]

Topp's paper provided vindication for Sheffey, who was upset by a CBS television report on the proposed canal that, in his view, minimized the fouling and ballasting means of marine species transport that had been occurring since the Panama Canal's opening in 1914.[99] But Rubinoff countered that such modes of dispersal rarely contained enough individuals of any given species to allow

CHAPTER 6

Avoiding an Elastic Collision with Knowledge

BY THE START OF the Canal Study Commission's final full fiscal year, July 1, 1969, many diplomatic and technoscientific difficulties had slowed down the course of determining the sea-level canal's feasibility. The collapse of the treaty reform process, the delayed nuclear test shots, Panamanian and Colombian opposition to the field surveys, the diversion of surveying equipment to Vietnam, the discovery of unstable clay shales along the preferred nuclear route, and the annoying calls by marine scientists for deeper bioenvironmental studies had all undermined the optimistic assumption of the 1964 enabling legislation, that a decision could be rendered within three years. And yet not once but twice, the commission convinced Congress to extend its deadline and funding, a sign of the nuclear seaway's powerful allure as a technological solution—or, at the very least, of most legislators' faith in the commission's ability to adjudicate on such a momentous project.

The group's leaders were by no means paragons of objectivity and transparency; they embarked on their mission with preconceived notions about the value of employing PNEs for canal construction, and they sat on geological intelligence of significance to future treaty negotiators, an issue that extended beyond their mandate. But at the same time, the commission was much more flexible than the federal agencies with which it worked to demonstrate the feasibility of peaceful nuclear excavation, the Atomic Energy Commission and the Army Corps of Engineers. The five commissioners and their executive director ultimately accepted that they could not overlook the technical nor public relations problems of peaceful nuclear excavation, despite the enormous cost savings it appeared to offer.

This chapter examines the CSC's denouement, as its members figured out how to frame the final comprehensive report to the president in light of two unresolved issues: PNE feasibility and marine ecological effects. The first point consumed their attention throughout their existence, and one of their last related questions involved whether to support the AEC's fourth planned canal cratering experiment, Project Sturtevant. The second matter, about the biological, nonhuman consequences of joining the oceans without a freshwater barrier, absorbed an increasing amount of their time during the second half of their tenure, though they had received a heads-up early on from a perceptive diplomatic consultant.

The pressure the commissioners encountered from 1968 to 1970 to fund preliminary studies of nonradiological ecological effects happened to coincide with the rise of the modern environmental movement. Marine biologist Ira Rubinoff originally sought not necessarily to stop the project but rather to tap into the CSC's pool of funds, though other members of the scientific community disagreed. Despite the wide range of views expressed by researchers, the commissioners perceived all who predicted potential canal-induced extinctions as alarmists. The CSC leadership conflated them with the students and activists in the news raising their voices against the environmental costs of modern society.

Nineteen sixty-nine became a particularly important year for the rise of modern environmental advocacy. On January 28, an oil-drilling platform off the coast of Santa Barbara exploded, coating dozens of miles of the Southern California coastline with oil. Later that summer, hazardous waste in Ohio's Cuyahoga River caught on fire. Grassroots groups built on the momentum of earlier citizen and scientific protests against radioactive fallout to call for stronger regulations against industrial pollution and rampant development.[1] Congress responded, as did President Nixon; he approved the National Environmental Policy Act as his first official act of the decade of the seventies.[2]

Like Nixon, the leaders of the CSC cared little for environmental causes, but they recognized the public relations value of supporting them in at least limited ways.[3] Despite the constraints they faced during their final two years, the CSC did expand the bioenvironmental studies to go beyond the public health effects of radiation in the isthmian environment, and it commissioned the National Academy of Sciences to produce a comprehensive ecological research agenda to precede the seaway's construction.

The commissioners' final assessment regarding the risk of adverse ecological consequences—that it appeared acceptable—elicited intense criticism from scientists, environmentalists, and antienvironmentalists seeking to preserve the Canal Zone status quo. *Science* published a harsh assessment of the academy's failure to influence the commission, and a marine researcher accused the CSC's executive director of "having an elastic collision with knowledge," judgments that have influenced analysts ever since.[4]

Yet the terrible publicity regarding the CSC's assessment of the sea-level canal's nonradiological effects bears further scrutiny. Not only did it overlook important extenuating factors regarding the state of marine invasion ecology in the late 1960s, it also obscured the remarkable transformation the commission members experienced on the topic of peaceful nuclear excavation.

The National Academy and the Canal Question

The idea that the U.S. National Academy of Sciences (NAS) should play a role in the sea-level canal debate had gained prominence as marine researchers debated both among themselves and with the CSC about the potential harm wrought by colonizing species. Rumors that the NAS might appoint an investigative committee circulated at oceanographic and zoological conferences following the publication of Rubinoff's pivotal 1968 *Science* article.[5]

Yet NAS involvement was not a foregone conclusion. The academy was established in the 1860s to provide the government with expert advice, which during its first century pertained mostly to the physical sciences. As the 1960s-era environmental movement unfolded, the academy faced pressure to admit more ecologists, especially in the wake of its advice reinforcing the status quo on chemical pesticides.[6] LaMont Cole was the most outspoken about urging the organization to think ecologically. He vented to Ernst Mayr that a 1966 NAS report on the plant sciences had recommended allocating $1.5 billion over the next decade for molecular research "but nothing for ecology per se, and not a penny for acquiring study areas or preserving natural areas . . . It is this sort of thing that requires us to see that Congress and others know where to go for competent ecological advice, and that they learn to recognize when ecological problems are involved."[7] But for Mayr the problem went much deeper: "It is not ecology which is getting the short end of the stick, it is the life sciences."[8] Cole took Mayr's reproach in stride, concluding, "I'll try to be discreet and I wish you success boring from within."[9]

Like a burrowing shipworm, Mayr did weaken academy resistance to entering the fray. In an April 1968 letter to the NAS president, physicist Frederick Seitz, Mayr invoked the elite private organization as the needed corrective to governmental mismanagement: "The forthcoming man-made mixing of the water masses, faunas and floras of the two oceans, is perhaps the most gigantic biological experiment ever undertaken by man. Considering the magnitude of the research required prior to the opening of this canal, it is deplorable how little thought and attention has been given to this problem up to now." A committee backed by "the prestige of the National Academy" would be able to develop a comprehensive research plan and raise funds to implement it.[10] Seitz invited Mayr to chair an ad hoc committee to address research on "ecological unbalances" for just one day, in July 1968.[11] After further nudging, Seitz agreed, in May 1969, to establish a substantial committee, to be funded by the CSC.[12]

Getting to that point took a lot of work. Ripley's team at the Smithsonian spent weeks with CSC personnel to craft an official letter from the CSC requesting the academy's assistance.[13] The document proposed an intricate set of goals. Because the CSC lacked the time and funds for nonradiological research, it recommended that the Smithsonian develop and execute a coordinated program of ecological studies, in concert with the advice of academy-designated experts. Academy approval of the Smithsonian role would, everyone hoped, spur Congress to appropriate the requisite funds.[14]

CSC Chairman Robert B. Anderson sent the letter to Seitz on December 6, 1968.[15] Because the letter did not disclose that the commission had already invited the Battelle research organization to submit a $214,000 proposal to evaluate the sea-level canal's nonradiological biological impacts using preexisting data, the discovery thereof caused embarrassment and irritation.[16] Yet despite questions about duplication, the agreement proceeded.[17] Too much was at stake to stop: the Smithsonian leadership had invested years campaigning for funds; the commission sought to disarm its ecologically oriented critics; and for the academy, the eighteen-month-long project provided an opportunity to expand its influence, as well as $50,000 during a tight period for federally funded research (the current equivalent would be over $365,000).[18]

Amid the last-minute negotiations, a devastating critique of the CSC appeared in the journal *Science* on January 10, 1969. The article painted an unflattering portrait of the commission's executive director: "Sheffey does not view the potential environmental consequences of a canal as particularly serious. 'The possibilities of any serious disruptions to nature are very remote, and the potential threat to biota is so insignificant that it doesn't merit spending a lot of money on it.' " The article also quoted him as calling out the Smithsonian scientists for "taking an alarmist view to attract attention."[19] Sheffey protested that the journalist had misquoted him and had failed to convey CSC efforts to overcome its lack of funds for ecological research.[20]

Three days later, on January 13, 1969, the commission members debated the wisdom of funding the $214,000 program outlined by Battelle, which featured the development of a computer model to determine what would happen to marine species following the opening of an artificial waterway at sea level. Given the $3 million already devoted to radiological studies, would such data really affect the government's decision about whether to build the new canal? No, answered Anderson, "but we would be able to tell the conservationists and other people who are concerned with this sort of thing what the probabilities were and whether there were steps such as fish nets that might or might not be employed" to prevent harmful biological effects. Four years after the canal treaty negotiator John Irwin had made a similar suggestion, Anderson admitted, "I don't think we can afford to say that we have studied everything except the marine life disruption."[21]

The other members agreed on the public relations value of funding the faunal-mixing model. Still stinging from his public reprimand, Sheffey pointed out, "We are already getting articles in the *New York Times* and the science magazines that we are not doing enough on it, and these people are much encouraged that we are contemplating doing it now." Commissioner Hill, the hard-nosed water resources management expert, did not conceal his contempt for those asking about marine effects. The commission should have answers ready, he argued, because otherwise "there would be no restriction on the wild eyed nature groups in causing trouble," and because "it gives us a little leverage to combat the uninhibited conservationists and the nature lovers—as the British call them, the birds and bees boys." Hill conflated the scientific sea-level canal authors with political environmentalists, even though the scientists argued among themselves about whether the megaproject posed a threat to be opposed or a research opportunity to be seized.[22]

The commissioners finally agreed to authorize the $214,000 expenditure, and to go up to $250,000 if need be, while dismissing Smithsonian official Sidney Galler's criticism of the Battelle proposal as "a little professional jealously among biologists and agencies." The commissioners also approved the $50,000 contract with the National Academy. As Sheffey explained, "To distinguish the two, the Battelle program is an action program for a product and an answer at the end of it. The National Academy program is [a long-term research] one to be executed only if the canal is built."[23]

With the "birds and the bees boys" fire put out for the time being, the Anderson Commission turned to the problem of handling its strident nuclear consultants. Sheffey had started drafting the final feasibility report by emphasizing that much more research and development were needed to determine whether nuclear excavation could be used to dig a new canal. However, his "degree of negativism" appalled John Kelly and other AEC officials. Once again, Commissioner Hill did not mince words in his response: "Sometimes I think we went beyond what we should have done to help the experimental [nuclear excavation] program. But now we are up to the point where these statements that have been made are in conflict with what we know to be—and I say 'know' from a technical standpoint—either to be improper or that they clearly are not going to be resolved." Hill had recently visited Livermore to discuss the issue of the Route 17 clay shales with Plowshare scientists, who had conducted follow-up chemical experiments regarding the soil mechanics in question. He left disappointed that "a lot of this was wishful thinking and not cold blooded technical analysis." Therefore, the report *should* exude a negative tone "because the time is coming when we are deceiving people and telling them something we know can't be met."[24]

Yet another complex development facing the commission pertained to the Panamanian military government's desire to resume the treaty reform process. Although the negotiations had been suspended since the 1967 leak of the draft treaties, Anderson met in his capacity as lead treaty negotiator with Panama's dictator, General Omar Torrijos, in September 1969. Anderson explained to the canal study commissioners that Torrijos was no ordinary strongman, but rather one who saw himself as having deposed an elitist, unrepresentative government that had been in place for sixty years: "What he is trying to do, he says, is bring about more democracy by means of greater participation by the average citizen and less participation by the oligarchy of families. . . . I think we have to be somewhat tolerant of the other man's view of democracy and not always try to export our own brand."[25] The Panama Review Group had already recommended to President Nixon that future canal negotiations seek definitive rights to a sea-level canal in Panama, though the treaty negotiations did not resume until after the CSC ruled out nuclear methods.[26]

Because the Anderson Commission and other insiders had avoided making the clay shales discovery known to Panama, a major discussion ensued at its final 1969 meeting, in October, about whether to endorse the next scheduled nuclear cratering shot, a 170-kiloton explosive known as Sturtevant. As in the case of Cabriolet and the other canal-oriented Plowshare tests conducted at the Nevada Test Site, strong arguments existed for not risking a violation of the Limited Nuclear Test Ban Treaty, especially after the 1968 Schooner debacle. Now even more was at stake for the U.S. with respect to its bargaining position with Panama. Commissioner Fields argued that the test should take place in order to maintain leverage and to support the Plowshare endeavor: "The one route that stands out, I think, is really [Colombia's] Route 25 for [nuclear] feasibility. If you don't test now and it becomes apparent to the world that you are not going to, there is no reason why you should continue with nuclear excavation technique, so you cut it off. And then you are dealing with Panama and you are at their mercy."[27] He also sought evidence in favor of combining Routes 17 and 25 across the Darién, a proposal recently advanced by Panamanian and Colombian officials—though they rejected nuclear methods.[28] Fields's colleagues did not share his eagerness to pursue the Sturtevant test.

By the time the commission held its next meeting, on January 22, 1970, another portentous political event had occurred. On the first day of the new decade, President Nixon signed the National Environmental Policy Act (NEPA), a six-page statute requiring federal agencies to consider environmental consequences and open up environmental project decision-making to greater public involvement. Congress had passed the law with little debate, and Nixon used the signing ceremony to speak only in general terms about reviving "a productive harmony between man and nature."[29] Not until months later did the law's revolutionary implications for development proposals using federal funds or lands became shockingly clear.

Two weeks after Nixon heralded the opening of the "environmental decade," an AEC commissioner, Theos Thompson, assured the attendees of a Las Vegas symposium on PNEs that the new law complemented rather than contradicted its mission: "The Act creates a three man Council on Environmental Quality within the White House to recommend environmental policies to the President and it requires all Federal agencies to take into account the environmental impact of all actions they propose. Of course as you know, the AEC has been doing this since its establishment." As evidence, Thompson referenced congressional hearings held by the Joint Committee on Atomic Energy in 1959 on the biological and environmental effects of nuclear war, as well as the Project Chariot Alaskan ecological study. In his view, the Chariot harbor plan, like Plowshare's other geoengineering projects, epitomized rather than contravened the NEPA mandate to "create and maintain the conditions under which man and nature can exist in productive harmony." For Thompson, furthermore, environmentalists' rejection of American ideals of progress and technological change constituted a possibly greater danger than threats to the environment per se.[30] As would soon become evident, such attitudes echoed those of the CSC members.

Thompson concluded his speech by praising Plowshare's great potential "to improve the quality of our environment, enrich our understanding of the ecology and natural resources and enhance our efforts to achieve a productive harmony between man and nature"—in other words, to meet the broad goals of the NEPA statute. However, his claim that the AEC had always paid careful attention to the environmental consequences of its mission did not speak to earlier scientific debates over thermal pollution, ocean dumping of radioactive waste, and other threats posed by nuclear plants to nonhuman organisms.[31] The closing remarks that a Plowshare official, Livermore physicist Glenn Werth, made at the 1970 Las Vegas conference reiterated the failure of the atomic bureaucracy to reckon with the values embodied by NEPA and the modern environmental movement. He repeated the old arguments about the economic savings of nuclear canal construction while also proposing to use nuclear energy to clear smog, tap non–fossil fuel energy sources, and preserve scenic landscapes by storing wastes underground. "If we have the foresight to set aside emotional irrationalities," Werth asserted, "we can move forward using nuclear energy and nuclear explosions to improve our environment."[32]

The ability of AEC officials to continue moving forward with their visions of harnessing atoms for peace, however, soon faced formidable challenges. In 1971 "an upstart environmental organization," the Chesapeake Bay Foundation, used NEPA to contest the AEC's process for licensing a new nuclear power plant in Calvert Cliffs, Maryland. The court case produced a landmark decision that transformed the commercial nuclear industry and warned all federal agencies to comply with NEPA's procedural requirements.[33] The Calvert Cliffs verdict forced the AEC to suspend reactor construction for seventeen months as it filed the appropriate impact statements, leading a legal scholar to ask in late 1972 whether NEPA constituted an "environmentalist Magna Carta" or a "coup de grâce" to the atomic energy establishment.[34]

In the meantime, as publicity in the early months of 1970 intensified for a series of Earth Day events on April 22, the Nixon administration allowed the meteorological window for conducting the Sturtevant test to pass. Yet Plowshare proponents continued to try to convince the canal study commissioners that Route 17's clay shale soils could be safely excavated with nuclear devices. Having conferred with Soviet nuclear scientists during recent trips to Moscow and Vienna, Kelly reported good news at the commission's March 1970 meeting. While the Soviets had not responded to his team's questions about clay shales on their home territory, at the subsequent International Atomic Energy Agency conference on PNEs, "they showed up in Vienna with reports and said they did not understand the U.S. concern on this problem of slope stability in clay shale material because they had a large amount of experience." The Soviet physicists presented evidence that "slopes produced by explosions in clay shales were tremendously more stable than those produced by more conventional excavation techniques." Kelly considered their explanation reasonable: "The explosion squeezes the water out of it [the clay]. It becomes pretty hard and impervious and it does tend to stand better than if you just scrape pieces off." The commissioners took the claims with a grain of salt. "In spite of the optimism expressed by Mr. Kelly of the Russian experience," asserted Hill, no nuclear explosions had occurred long enough in the past to provide long-term evidence of erosion-resistant slopes.[35] Four months later, his patience fully eroded, Hill delivered a stern rebuke to his fellow commissioners for having "played along with" Kelly's optimistic expectations about PNEs for far too long.[36]

By the spring of 1970, the commission was equally fed up with the scientists seeking CSC funds to explicate the seaway's potential consequences for marine life on either side of the isthmus. At the March meeting, Sheffey admitted that though the Battelle contractors were still at work, he had a good idea of what their result would be: they would conclude that the discernable threats, such as invasions of coral-eating starfish, were minimal. "The alarmists, like Dr. Briggs and Dr. Rubinoff," on the other hand, "are going to continue to scream that thousands of species are threatened, but they can't prove it." Replied Eisenhower, "Every time civilization has made a move, opening the West or anything else, it has changed nature." Hill mentioned an article he had read about activists in Death Valley, California, who were trying to stop a development project based on the discovery of "some diminutive fish . . . that roots in the mud in one creek out there," a probable reference to pupfish conservation efforts initiated decades earlier by the ichthyologist Carl Hubbs.[37] Despite the extreme rarity of the aquatic species, efforts to list it as endangered did not impress Hill: "What difference does it make? None," he answered.[38]

One month before the highly anticipated Earth Day teach-in events of April 1970, the commission members agreed that the final report should include a statement that the faunal-exchange problem should not be allowed to block a decision in favor of a sea-level canal. Added one of the engineering consultants, "The one thing you can do to get them [alarmist scientists] off your back is going to come from the National Academy study, which is a program of investigations that you should pursue in the years to come to answer all these questions. As long as you offer them an opportunity to study, they are happy."[39] It was an ironic declaration given the role that scientific evidence had played in undermining the Chariot project in Alaska.

The Committee on Ecological Research for the Interoceanic Canal

In the meantime, the *Science* exposé that had portrayed Sheffey as antipathetic to ecological questions generated more press. The Nobel prizewinner Joshua

Lederberg devoted his weekly *Washington Post* science column to the proliferation of "mega-experiments" with global yet unknown environmental effects, as epitomized by nuclear testing and the chemical pesticide DDT, and to the concomitant political need for ecological research. Predicting the sea-level canal's consequences for oceanic life required a range of approaches, old and new: "As a molecular biologist interested in evolution," he wrote, "I would at least insist that large samples of present marine life at various stations be carefully frozen for later examination." Lederberg urged the CSC not to allow diplomatic, engineering, military, political, and financial issues to bury the scientific ones.[40]

That was precisely the point of the contract with the National Academy establishing CERIC, the Committee on Ecological Research for the Interoceanic Canal. In the spring of 1969, Mayr, who had agreed to chair the committee, met with several advisors, including Wallen and Galler of the Smithsonian, John Wolfe of the AEC, and marine and hydrological scientists associated with the National Science Foundation, U.S. Geological Survey, and Woods Hole Oceanographic Institution, to discuss the committee's scope and membership.[41] They identified dozens of research questions pertaining to marine biota, the biology of species dispersal, and the physical parameters governing water movement through the canal as well as the conditions on either side of the isthmus.[42] Above all, stressed Mayr, "at this stage it is most important to counteract the widespread impression that the potentiality of 'fall-out' is the only, or at least major, problem of the Sea-level Canal."[43]

Following a months-long process to pin down the membership, CERIC's first formal meeting took place in July 1969 at the academy's stately headquarters in Washington, D.C. The ten members seem to have been chosen, primarily by organizational affiliation, by NAS executives, with input from Mayr. While he did not succeed in having Rubinoff appointed as the Smithsonian representative (a position filled by David Challinor, who succeeded Galler as the assistant secretary for science in 1971), Mayr did gain representation for two Caribbean university marine biology programs.[44]

However, left off the list was an institution with a strong stake in the matter, the University of Miami's Institute of Marine Science. The school employed Gilbert Voss, Frederick Bayer, and C. Richard Robins, one of the teams subcontracted by Battelle, the private organization that had won out over the Smithsonian in the 1965 competition for AEC funds. The Miami researchers had initiated transisthmian oceanographic trawls long before receiving any of the canal study funds, and they were still working on their report for Battelle when CERIC came into being.[45] As Challinor later testified before Congress,

CERIC's dismissive attitude of the Miami group seemed to relate to their sampling techniques and their assessment that most species exchange through a sea-level canal would likely be minimal and noncatastrophic.[46] Routine academic elitism and rivalry probably also played a part.

Mayr wanted an eminent ecologist to serve on the committee and convinced his younger colleague Edward O. Wilson to do so. As codeveloper of the innovative theory of island biogeography, the insect systematist aimed to revolutionize ecology by linking it with genetics and biogeography.[47] Wilson had also participated in a foundational 1964 symposium at the Asilomar Conference Grounds in California on the genetics of colonizing species, presenting on the invasive—or as it was then called, imported—South American fire ant.[48]

At CERIC's first workshop, held in August 1969 in Woods Hole, Massachusetts, Mayr explained the committee would proceed on the assumption that the sea-level canal would actually be built. The CSC allowed him to share its major recommendations, which would not be publicized for another eight months: conventional, not nuclear, explosives would be used; the new waterway would be built in Panama along Route 10, ten miles west of the existing canal; and construction would begin around 1982, allowing researchers just over a decade to complete the baseline studies to be outlined by CERIC.[49]

The group spent its first meeting discussing both the optimal scope of the research program and specific hypotheses about dispersal and colonization. Wilson's presentation, for instance, noted the repeated failure of intentional species introductions, and the need for field and laboratory experiments to determine the qualities of successful colonizers. As for the range of research to be recommended, some participants expressed fear that not enough qualified experts existed to carry out a large-scale research program; others sought to "think and act big" to impress Congress in order to attain the funds, which would then help train the requisite taxonomists, biogeographers, and ecologists.[50]

The CERIC membership met three additional times, in October 1969, in January 1970 for a week in Panama, and in April 1970, but the bulk of its work was handled by Staff Officer Gerald Bakus, a coral reef specialist who relocated from California to Washington to take on the immense job.[51] To assess the current state of knowledge on selected Central American marine biota and how human effects might change them, Bakus contacted nearly two hundred specialists with exhaustive requests for information.[52] One informant complained he would need six months and "at least $10,000 plus perquisites" to do the job.[53] Even getting the CERIC members to draft and comment on their assigned report sections was difficult.[54] On the other hand, the Miami researcher Voss later complained his team was never contacted, despite the years they spent surveying isthmian waters on the R/V *Pillsbury* and searching the literature in obscure libraries.[55] Bakus might have assumed the Battelle-funded team members were not willing to share their data, signifying the mutual mistrust of the two scientific groups contracted to assist the CSC.[56]

Despite their different ecological predictions, both groups agreed the sea-level canal should contain a precautionary barrier incorporating physical, thermal, or freshwater obstacles to marine migrants. Mayr had made a point of asking the CSC to investigate the feasibility of such structures, and he drew on a recent event that seemed sure to capture their attention, the July 1969 moon landing: "It is a situation somewhat analogous to the potential risk of astronauts bringing a highly dangerous pathogen from the moon to the earth. Even if such a danger had only a very small probability, it should be avoided at all possible cost."[57]

Another seemingly ideal opportunity to convince the CSC to recommend a precautionary approach occurred in December 1969. John Briggs, the most conservation-oriented of the sea-level canal authors, organized a symposium at the annual meeting of the American Association for the Advancement of Science (AAAS) in Boston that featured presentations by Sheffey, Voss, Topp, and other sea-level canal authors. Yet to Bakus's surprise, the symposium speakers "almost unanimously agreed that there would be only slight biological effects from interoceanic dispersal."[58] Rubinoff attributed the optimism to ignorance of basic concepts such as competitive exclusion, the principle that two species vying for the same resource cannot coexist.[59] Bakus concluded that many speakers overlooked evolutionary mechanisms of species dispersal, and attributed some of the disagreements about the likely patterns of migration, colonization, and competition to diverse disciplinary outlooks.[60] Almost a decade later, the malacologist Geerat Vermeij made the related point that "the direction and the magnitude of any biotic migration through a Central American sea-level canal are likely to differ among taxonomic groups as well as among organisms from different communities," hence the need for "greater insight into the properties of natural history of individual organisms and species" as opposed to "studies on abstract group measures such as diversity or community stability."[61] Vermeij spoke to a fundamental conundrum of postwar ecology—despite the growing interest in developing predictive models, the low prestige of old-fashioned naturalist-based disciplines hampered efforts to acquire basic yet crucial information.[62]

The AAAS scientific speakers did acknowledge one specific danger, the crown-of-thorns starfish. The topic was on the minds of marine biologists because the

Indo-Pacific *Acanthaster planci* was preying on the coral of Australia's Great Barrier Reef, and the first major paper on the topic had just been published by one of the sea-level canal authors, Richard Chesher.[63] In his presentation, Voss stated that a little-known eastern Pacific species, *Acanthaster ellisi*, appeared almost identical to *A. planci*; a Central American sea-level canal might thus enable *A. ellisi* to prey on Caribbean corals.[64] A few months later, STRI's coral reef ecologist, Peter Glynn, confirmed the first *Acanthaster* outbreak west of the Gulf of Panama, leading Briggs to predict, "If the crown of thorns got into the Atlantic, there would be a very great risk of damage all the way from the Florida Keys to Rio de Janeiro."[65]

The other dramatic species that captured public attention was the eastern Pacific yellow-bellied sea snake. To quantify the invasion potential of a specific organism, Rubinoff had shifted his attention from gobies to the physiology and ecology of *Pelamis platurus*, a venomous organism that periodically appeared in large numbers along Panama's Pacific shores. By offering the snake to potential Atlantic predators, Rubinoff and his collaborators sought to predict how fast it might colonize the Caribbean Sea and adjacent Atlantic Ocean. Their experiments indicated that Atlantic predators were much more likely to attack the serpents and die than Pacific carnivores, and thus following the construction of an unobstructed channel, natural selection would favor predatory fishes disinclined to attack sea snakes.[66] Due to its ability to drift and feed at the surface, *Pelamis* might not only invade the Caribbean and Atlantic but also ride the Gulf Stream all the way to the English Channel.[67]

For Sheffey, it made no sense to focus on an organism that did not seem to pose problems in its native habitat and that had not already transited the existing canal: "I have been swimming many times on the Pacific side of Panama, in Hawaii and in the South China Sea off Vietnam. Until I met Rubinoff I had never heard of the sea snake, and to this day I have been unable to find anyone who has ever heard of anyone being bitten by one."[68] But for Rubinoff, the Pacific sea snake constituted a "conspicuous example" of just one of the many species capable of being transmitted in large, reproducible numbers through a sea-level canal, and one that could help people conceive of the problem of invasive species. As he acknowledged to a colleague, "there are many less conspicuous organisms which may cause greater economic problems to commercial enterprises."[69] These points would generate much attention later in the 1970s, when the sea-level canal proposal reemerged in a very different political context.

An Elastic Collision

After an intense year, by the spring of 1970 CERIC's report was near completion. The 231-page document outlined detailed ecological, systematic, and oceanographic studies to be conducted prior to seaway construction, including the establishment of a faunal bank to store samples of organismal tissues for genetic analysis. The preliminary research program would require at least ten years, an initial capital outlay of $4 million, and annual budgets of about $2 million, to be paid for by "a new system" of environmental cost accounting borne by the waterway's users. The committee also called for a precautionary faunal barrier, concluding, "The construction of a sea-level canal in Panama is a gigantic experiment with natural ecosystems whose consequences are unforeseeable."[70]

But most of these points wound up being overshadowed by a front-page *Washington Post* article that proclaimed "A-Canal Dealt Blow." Appearing nine days before the hyped Earth Day events of April 22, 1970, the article quoted Mayr on his personal opposition to nuclear excavation, and implied that the academy biologists possessed far more power than in actuality by leading with the statement, "The dream of a future sea-level Atlantic-Pacific canal blasted out cheaply by nuclear explosives has been dealt a severe blow—maybe a fatal one—by a group of biological advisers to the canal study commission."[71] The portrayal distressed Bakus and the academy leadership, but Mayr conceded only that they should have tape-recorded the interview: "None of the recommendations of our committee were 'leaked' to the press and I carefully refrained from any value judgments. . . . All this will blow over in a couple days. So cheer up!"[72]

Far from blowing over, the situation got Congress's attention, though not the kind everyone had hoped for. Representative Dan Flood used Mayr's words to denounce the wastefulness of the CSC and to reiterate his long-standing argument for expanding the existing canal. The previous autumn, upon learning of the forthcoming AAAS meeting, Flood had contacted the biogeographer Briggs to urge him and other biologists to send him copies of their articles (and to write to their senators and representatives that the U.S. Constitution granted Congress, not the president, the power to dispose of U.S. territory).[73] As was his custom regarding any publicity threatening U.S. interests in Panama, Flood entered the canal ecology articles into the *Congressional Record*, providing a new source of ammunition to his colleagues on the political right who opposed any change in U.S.-Panama relations.[74]

In the meantime, the CSC and CERIC locked horns over the focus of the final CERIC draft. Baffled by the "alarmist viewpoint" expressed despite the

AAAS session's optimistic assessment, Sheffey urged a more balanced discussion of both the potential dangers and mitigating factors.[75] But CERIC members pushed back, especially Scripps cirripedologist William Newman, for whom recommending a biotic barrier was no more alarmist than suggesting a fish ladder for a dam.[76] Newman later publicized Sheffey's letter to the group and accused him of having had "an elastic collision with knowledge in his argument that the Crown-of-Thorns starfish (and, therefore, presumably many other organisms of which we know little or nothing) already would have established itself in the Caribbean were conditions there favorable for it."[77]

Newman's exposé appeared in a volume titled "The Panamic Biota: Some Observations Prior to a Sea-Level Canal," which grew out of a March 1970 conference organized by invertebrate zoology curator Meredith Jones of the Smithsonian Museum of Natural History. Along with essays by other sea-level canal authors was one by the University of Miami biologist Voss blasting CERIC for disrespecting his team's hard work. Their 480-page report featured marine isthmian species inventories compiled from literature reviews, trawling cruises, and interviews with biological oceanographers.[78] While Battelle funded most of the analytical phase, for the field operations the researchers had scrambled for support from other sources.[79] On these data the authors based their controversial conclusion that no "valid biological reason" appeared to exist for opposing a nonnuclear-excavated waterway, especially not "if certain safeguards are built in."[80] In turn, Battelle used the Miami report to inform a separate report to the CSC that mathematically modeled the potential transport of water, chemicals, sediment, and planktonic organisms between the oceans, concluding, "It is highly improbable that blue-water species like the sea snake and the crown-of-thorns starfish could get through the canal except under the most unusual circumstances."[81] By specifying "blue-water" organisms of the pelagic zone, the Battelle contractors overlooked numerous species of the deeper demersal and benthic zones near, and at, the bottom of the sea.

Such distinctions were lost on the CSC members and engineering consultants. Meeting in July 1970, they scorned CERIC's argument for a thermal species barrier, an expensive method of preventing species exchange via hot water. Nor did the commission appreciate CERIC's insistence on making value judgments regarding the risks at hand. Rather than simply recommending studies to be conducted, the academy group argued that the dangers of marine species exchange justified the installation of biotic barriers. By contrast, as Sheffey had predicted, the Battelle group fulfilled its narrow mandate of providing advice, which happened to fit the commission's view that the marine ecological risks of building the seaway appeared tolerable.[82]

To reconcile the two groups' different views in the final report, Sheffey asked the commissioners for policy guidance, which revealed their inability to grasp the rising political influence of the modern environmental movement. Stated Commissioner Hill, "If we had to go to a decision tomorrow, I would say forget the ecological hazards, that they are minimal; but we don't have to make a decision tomorrow, so maybe we can get something for the fellow who has to make the decision. . . . Until this group of ecologists and wave of hysteria dissipates, you have to live with it." Sheffey also asked for advice about whether to include "the ecology discussion" in a separate chapter or as part of the longer chapter on technical and financial considerations. Eisenhower's recommendation that he need not worry about the shortness of the ecological chapter would soon come back to haunt the group.[83]

While the commission members favorably interpreted the lack of data on the potential disastrous effects of seaway-induced biological invasions, they did not do so for PNEs. In fact, they spent one of their last meetings grappling with their own rapid evolution on the issue. "Really we were created with the idea that we were going to dig a nuclear canal. This gets overlooked," stated Chairman Anderson at the twenty-eighth meeting in July 1970. "In your narrative form," he told Sheffey, "let's try to recreate a part of the [pronuclear] atmosphere . . . [of 1964–65]. I must say I was part of it." Replied Eisenhower and Storey, respectively, "I was too" and "I was strong for it." In his blunt manner, Hill summarized the sea change that had washed over them during the past five years: "There was no expectancy when we started out that we were going to find that Route 17 was an impossible situation. . . . all of the work on 17 and most of the work on 25 was to demonstrate the feasibility of nuclear excavation. . . . I think the [nuclear excavation] chapter should describe that and what we did to prove it, and then it came about that we proved the opposite of what we expected."[84]

The fifth commissioner, Fields, expanded on Hill's assessment by addressing fundamental differences between the science and engineering of nuclear excavation. Route 17 had appeared optimal to the Livermore physicists, the developers of the nuclear cratering technology. But they failed to anticipate a key issue due to a lack of ground truthing: "They have always had the problem in the atomic energy field when you move from the basic science into the engineering." It was the engineering studies that exposed the clay shale permeating much of the preferred Panamanian route. And yet the AEC and Livermore scientists *still* believed they could overcome the clay shale conundrum—so much so that they accepted the claims of their Soviet counterparts over the American engineers. "But you find that the better of the engineering people will be against them, that in the foreseeable future you won't like this clay shale. So you have the scientists and the engineers against each other." Could the commissioners settle such differences? No, concluded Fields.[85]

The CSC delivered its 1,074-page tome to President Nixon on December 1, 1970, eight anticlimactic months after the revelations that technical uncertainties and international skepticism had rendered nuclear excavation unfeasible, and two weeks after insiders imparted the key recommendation to the media.[86] Most of the volume presented military and economic justifications for conventional construction of a $2.88 billion, 550-foot-wide sea-level channel with tidal gates along the Route 10 site just west of the Canal Zone.[87] The commissioners had rejected the idea of adding a third, wider lane of locks to the existing waterway because it would buy only twenty years before ships exceeded the new locks once again and would require pumping in massive quantities of seawater. Furthermore, even if the original canal were operated free of tolls, economies of scale would likely lead the shipping industry to divert the bulk of traffic to much larger ships by the turn of the twenty-first century.[88]

Although the commissioners considered the Route 10 location disadvantageous due to the need to acquire new lands, it was the shortest of the five routes considered and the only Panamanian one that would not interfere with the existing canal. Route 10 would also go through Gatun Lake, thereby preserving a freshwater obstacle to marine species exchange, as presented in a chapter that would have been unthinkable in 1965: "Environmental Considerations." Moreover, in accordance with NEPA, the appendices included brief environmental impact statements for the proposed routes.[89]

Yet in their accompanying classified letter to the president, the commissioners made no mention of environmental issues. Rather, they addressed the sea-level canal in the context of future treaty negotiations with Panama and the unquantifiable benefits of ensuring U.S control of a modernized isthmian canal system for several decades thereafter. The existing canal had contributed to U.S. national security and enabled the United States to influence the economic development of several Latin American countries, the transportation costs of many U.S. exports and imports, and the trade patterns of all nations using the canal. Though the astronomical costs of building a new sea-level waterway with conventional methods could not be recompensed quickly by tolls, the commissioners concluded, the "risk of financial loss" would be offset by maintaining U.S. control of the Panamanian transportation corridor. Accordingly, because "pressure upon the United States to abandon its position in Panama will continue," a new sea-level waterway would "offer an internationally acceptable justification for our continued presence in Panama." Negotiating "a generous treaty" to allow continued U.S. operation and defense of the existing canal, eventual construction of a wide and deep sea-level channel, and a long period of U.S. control thereafter for amortization "could combine to produce more tranquil relations into the foreseeable future."[90]

The U.S. ambassador to Panama, Robert Sayre, congratulated Anderson on the final, public report, describing the Panamanian reaction to it as positive. Although officials expressed concern about the additional lands that would have to be obtained, and the possibility of creating another artificial river with no bridges over it, the report's focus on the value of building the new waterway in Panama generated positive press. "Of course," acknowledged Sayre, that was not the whole story: "No one has read the foreign policy and defense annexes which may create some sparks when they become general knowledge."[91]

In the United States, the news that Lyndon Johnson's CSC had recommended building a new sea-level canal in Panama with ordinary bulldozers and dynamite unleashed a tidal wave of condemnation. Representative Flood asserted that "there was never any doubt that when the commission finished that would be their conclusion" and that the group "wasted bags of money" surveying remote routes in eastern Panama and Colombia that it knew could not be used for the new canal.[92] On the other side of the political spectrum, biologist and environmental activist Barry Commoner derided the canal scheme's monumental scale and the concept of peaceful nuclear applications: "Plowshare has been a $138 million exercise in futility. It has foundered in the environment."[93] Thirteen years after Plowshare's genesis, the idea that nuclear excavation had been doomed to failure was already becoming conventional wisdom.

The CSC report's skimpy analysis of environmental issues angered the CERIC and Smithsonian scientists who had lobbied on behalf of preliminary ecological research for so long. The environmental impact statements were repetitive and superficial, and the "Environmental Considerations" chapter constituted only four pages, almost half of which addressed the assessment of the Battelle Memorial Institute and associated University of Miami team, with only one sentence noting CERIC's work. While conceding it would be possible to install a temperature or salinity barrier "should future research indicate the need for a biotic barrier in addition to tidal gates," the commissioners concluded, "the risk of adverse ecological consequences stemming from construction and operation of a sea-level Isthmian canal appears to be acceptable."[94] In a year that had featured the first Earth Day celebration and mainstream coverage of the environmental movement's suspicion of technology and technocracy, such a conclusion seemed to confirm the growing reputation of engineers as "diligent destroyers."[95]

Science responded once again with a scathing critique. Subtitled "How the Academy's Voice Was Muted," it emphasized the CSC's privileging of the Battelle-Miami's upbeat analysis, and featured an academy-leadership-defying interview with Mayr that included the quote, "We said that great danger would result from building a sea-level canal, though we can't prove it. But they turned it around and said that, since we can't prove it, the danger is minimal."[96] After asking why the commission "largely ignored" CERIC's views, the author, science journalist Philip Boffey, argued that the academy was also to blame for allowing itself to be "mouse-trapped into a restricted role in which its voice was inevitably muted."[97] In other words, the academy leadership fell down on the job by not reserving the right for CERIC to recommend against canal construction, and by focusing too much on containing leaks rather than allowing CERIC members to communicate with the press in a timely fashion.

Boffey later published an important book about the National Academy of Sciences that blew the whistle on the close ties among many of its expert panels and special interests, but in this case he himself might have been too beholden to his informant. As Mayr acknowledged in his private correspondence, he had contacted the journalist in frustration over the academy's delay in releasing his committee's conclusions.[98] When the CSC publicized portions of its final report in advance of the December 1 deadline, Mayr had assured Rubinoff that once distributed, the CERIC report would "have far more authority than such statements in the press."[99] But the academy did not print enough copies, not even for the committee members.[100]

Boffey's other major criticism of the Anderson Commission, that it minimized the structure's ecological risks, was shared by other historical actors and analysts.[101] However, the report did call for further study of "a number of possible environmental problems" if the government decided to proceed with the project at some point in the future.[102] It is not fair to say that "the voices of concerned biologists, and CERIC's year-long study, were not heeded" by the CSC or that its final report "dismissed all the risks" raised by the sea-level canal authors, nor that "the controversy over the sea-level canal came to a close not because of scientific data or environmental risk, but because of politics" associated with the diplomatic transfer of the Canal Zone and waterway.[103] Such assessments reflect the viewpoints of specific historical actors, especially Mayr and the science reporter he contacted to produce the "How the Academy's Voice Was Muted" editorial.

Later in his career as a historian of biology, Mayr provided a robust endorsement of revisionist history that hinted at his own conflicting roles as actor and analyst: "Written histories, like science itself, are constantly in need of revision. Erroneous interpretations of an earlier author eventually become myths, accepted without question and carried forward from generation to generation. A particular endeavor of mine has been to expose and eliminate as many of these myths as possible—without, I hope, creating too many new ones."[104] As a participant in the sea-level canal controversy, Mayr used his journalistic access to shape a particular viewpoint, one that remains cited to this day. Yet by preserving his private correspondence for future scholars, Mayr the historian left the door open for deeper interpretations to emerge.

Conclusion

It is a great irony that the Atlantic-Pacific Interoceanic Canal Study Commission is now so closely associated with the "environmental considerations" of the proposed sea-level canal, or rather the lack thereof. The commission's work wound up highlighting the increasing importance of environmental criteria in public works planning. Throughout 1971, congressional opponents of renegotiating the U.S.-Panama treaty publicized the scientific questions about the dangers posed by otherwise-obscure marine organisms. As Flood demanded, "Why does the State Department ignore the marine ecological angle involved in constructing a saltwater channel between the oceans, which recognized scientists predict would result in infesting the Atlantic with the poisonous Pacific sea snake and the predatory crown-of-thorns starfish and have international repercussions?"[105] The International Affairs section of the Office of the Deputy Under Secretary of the Army received numerous queries on the CSC report, the overwhelming majority of which addressed the ecological effects of a sea-level canal. Consequently, State Department officials speculated that the Senate would focus "significant attention on the ecological aspects of the sea-level canal option in its consideration of any new treaty."[106]

The under secretary of state raised awareness of the issue among the highest levels of U.S. government. A classified National Security Council memorandum issued during the summer of 1971 asserted, "Greater attention must be focused on the question of the ecological impact of construction of a sea-level canal."[107] The lead author was none other than John Irwin, the former deputy treaty negotiator and New York Zoological Society trustee who had tried to convince the commission six years earlier not to neglect the marine ecological angle. Granted, the memo also asserted that defense and foreign policy goals no longer justified construction. The biologists' concerns did not shut the project down, but they did constitute another important rationale against it, especially in the context of a new era of statist environmental management.

The Canal Study commissioners navigated a difficult course during the five years of their tenure. Scientific discussions of the environmental risks posed by the sea-level canal did influence federal officials and policy makers in unexpected ways, and the CSC took a more nuanced approach to the question of environmental considerations than one of complete dismissal. Despite tensions and biases, its engineering agents and executive director worked for years with the Smithsonian and National Academy to increase the congressional appropriation for marine ecological research.

The CSC members did indeed fail to appreciate the problem of marine invasive species, and they failed to differentiate between scientists seeking financial support to study the potential effects of a proposed megaproject and activists seeking to shut the project down due to the presumed ecological risks. Nevertheless, the commission's blind spots about marine bioinvasions—a topic that would not gain widespread scientific and popular attention until later in the twentieth century—should not diminish how its leaders overcame their own technocratic embrace of Project Plowshare.

The CSC wound up caught between two eras, the high-modernist period of extreme faith in science and technology, and the new environmentalist age of increasing public suspicion of large-scale technological solutions. The U.S. National Environmental Policy Act of 1969 required federal decision makers to quantify the environmental costs of development, and thus imposed an enormous cultural change on entrenched bureaucratic agencies that had hitherto been incentivized to ignore such costs. Unlike the Army Corps of Engineers and the Atomic Energy Commission, the CSC was an ad hoc organization with no public constituency, whose authority diminished after Johnson's presidency. But it still had friends in Congress and commanded front-page newspaper coverage in 1970. Its ultimate refusal to make a recommendation for nuclear excavation technology, despite its enthusiasm just five years earlier, speaks to the rapid shift in values and expectations regarding environmental planning and policy-making.[108]

The Atlantic-Pacific Interoceanic Canal Study Commission was no model of democratic environmental decision-making, but neither was it as callous and hubristic as has been alleged. Its adaptively technocratic approach to determining the feasibility of the sea-level canal, and the proper role of nonanthropocentric ecological science therein, provides a clearer, more accurate view of the failure of an iconic high-modernist project to take root in the postwar era.

PART III

The Post-Panatomic Canal

CHAPTER 7

Optioning the Sea-Level Canal for the Energy Crisis

FROM THE ASHES OF the stillborn 1967 Panama Canal Treaties arose a most improbable phoenix, the 1977 agreement abrogating the Hay–Bunau-Varilla Treaty of 1903. Although the 1976 Republican U.S. presidential primaries had reiterated the extreme unpopularity of relinquishing control of the waterway and zone, Democratic president Jimmy Carter entered office on January 20, 1977, determined to complete the efforts of his predecessors to develop a more equitable relationship with Panama. Seeking a quick resolution to the venerable problem, he set his negotiators a tight deadline of six months. Although Carter prevailed, a fierce conservative backlash over yielding a source of great symbolic pride cost him, his senatorial allies, and the Democratic Party enormous reserves of political capital.[1]

The rancorous debate over the 1977 Torrijos-Carter Treaties featured the issues of sovereignty, national security, and economics, yet other concerns played subtle, intriguing roles over the course of the arduous negotiation and ratification processes—environmental ones. The pacts did, after all, come to fruition in the midst of the "environmental decade" and the energy crisis, and at the helm of a president with high standing among environmentalists, with a naval engineering background, and with a focus on reforming federal water project policy-making. Viewing this diplomatic and domestic watershed through an environmental lens reveals the rising yet uneven political influence of the environmentalist movement during the 1970s on megaproject infrastructural planning.

Although often assumed to have died once nuclear methods became untenable, the vision of a streamlined sea-level waterway revived in dramatic fashion in 1977–78 for a reason few could have foreseen in the previous decade: to confront an energy emergency. The 1968 discovery of America's largest oil field in Alaska, the 1969–73 debate over the proposed Trans-Alaska Pipeline, and the 1973 Organization of Petroleum Exporting Countries (OPEC) oil embargo led Alaska's Democratic senator to convince Carter that a sea-level canal would preclude the contentious construction of new transcontinental pipelines, among other benefits. Late in the treaty negotiations, during the summer of 1977, Carter compelled his negotiators to include an option for a future Panamanian sea-level canal, irritating many in both Panama and the United States.

The sea-level canal provision of the Panama Canal Treaty (the second of the two accords that negated the old one) has been described as a mere distraction and ploy to win an Alaskan ratification vote or two.[2] It is true that in contrast to the dominant role the idea played in the 1967 U.S.-Panama negotiations—when it was the subject of its own treaty—the seaway had become almost a nonissue for the 1970s-era mediators.[3] Its last-minute revival deserves to be taken more seriously for several reasons.

The postatomic rebirth of the sea-level waterway proposal challenges the idea that it died along with Project Plowshare in the early 1970s.[4] Moreover, it helps elucidate changes in U.S. statist environmental management during the pivotal environmental decade. The proposed megaproject shines light on the new opportunities and constraints of the National Environmental Policy Act of 1969, and on shifting attitudes about what constituted acceptable costs of implementing new forms of U.S. energy-security infrastructure following the 1973 oil shock.

The 1970s-era invocation of the interoceanic sea-level canal also provides a deeper look at Carter's enigmatic environmental record. He not only faced the challenge of balancing environmental goals with efforts to end the oil crisis and stimulate the economy but also sought to make the federal system of water project policy-making less wasteful and destructive. Early in his term, Carter tried to cancel plans for nineteen large-scale dams and reclamation projects that had already been authorized, an endeavor that alienated many members of Congress.[5]

While environmentalists commended the water projects "hit list," they detested the resurrection of the Panamanian sea-level canal. Building on the ecological and evolutionary insights of the marine biologists of the late 1960s, several prominent environmental organizations mobilized against the proposal, rather than welcoming it as an alternative to new terrestrial pipelines. The plan also contradicted Carter's own pioneering executive order recognizing the ecological risks of invasive species, and thus the 1970s phase of the seaway proposal illuminates changing public attitudes toward nonnative species, especially in terrestrial versus marine contexts.

Carter's advocacy of the nonnuclear sea-level canal speaks to the extraordinary challenges he faced as an environmentally oriented president steering the ship of state through the energy crisis. It fits the assessment that Carter sought to stimulate the economy while promoting environmental goals—a difficult balancing act that dissatisfied both his critics and supporters.[6] At the same time, Carter's insistence that the Panama Canal Treaty provide for the eventual construction of a sea-level waterway—like his determination to overturn the 1903 treaty itself—supports the idea that he adopted a trusteeship model of leadership that deemphasized the political costs to himself of pursuing unpopular courses of action.[7] Carter acted as a trustee of the public welfare by applying an engineering approach to solving certain problems as quickly and efficiently as possible, without consulting potential legislative allies, as in the cases of the canceled water projects and federal bureaucratic reorganization.[8] In other instances, his antipolitical trusteeship style featured a strong sense of morality.[9] The sea-level canal proposal, and the 1977 Panama Canal Treaty of which it was a part, exemplified Carter's technocratic and moralistic impulses to do what he considered right for the country, no matter the costs he incurred.

That is not to say that he ignored the importance of building political support for the cause of treaty reform; Carter knew the difficulties of ratifying the two pacts (the Canal Treaty transferred control of the waterway to Panama in 1999, and the Neutrality Treaty declared the canal open to vessels of all nations, while granting the United States the permanent right to defend it from any threat). Once Carter became aware of it, the sea-level canal option seemed to hold one of the keys to breaking down a few blocks in the wall of domestic resistance. If for President Johnson the seaway proposal had functioned as a tool to control both Panamanian nationalists and U.S. Zonian and defense interests, then for President Carter it seems to have served as a means of shoring up domestic support for the Panama Canal Treaties among an unlikely yet potentially influential coalition: Alaskan oil boosters, antipipeline environmental activists, and undecided senators seeking to ensure future cooperation with Panama on an exciting energy-security megaproject.

Ship of Joules

Newspaper editors do not often get the chance to use sixty-point font. But July 18, 1968, was no normal day for the state of Alaska. "ARCTIC OIL FIND IS HUGE" announced the *Anchorage Daily News*.[10] After years of exploratory drilling, and months of rumors, the Atlantic Richfield and Humble Oil companies confirmed they had discovered North America's largest field of subaqueous oil in Prudhoe Bay on the North Slope. The state government, which had only come into being in 1959, soon announced its intention to sell leases to corporations with the expectation of windfall taxes; the September 1969 auction of 450,000 acres netted over $900 million (over $6.5 billion today when adjusted for inflation).[11]

The discovery posed the huge question of how to get the oil riches to the rest of the nation's voracious energy consumers. A pipeline could be built through Canada to the northern tier of the midwestern United States, but that would require bilateral control and cooperation. For those seeking to keep it all in-state, a more expensive line could be constructed 800 miles south to the ice-free port of Valdez for transferring the oil to tanker ships. More dauntingly, a harbor could be excavated in Prudhoe Bay for East Coast–bound icebreaking supertankers to carve out the fabled Northwest Passage. So seriously was this last option taken that one of the major corporate players tested the 4,500-mile-long polar sea route by spending $54 million (the current equivalent of almost $380 million) to retrofit and send the S.S. *Manhattan* from Chester, Pennsylvania, to Point Barrow and back during the summer and fall of 1969.[12] In the meantime, Project Plowshare's unflagging advocate Edward Teller proclaimed at a Houston press conference that the Lawrence Radiation Laboratory at Livermore was ready to use PNEs to engineer an offshore atoll-style harbor, so long as the oil companies paid for it. That did not happen, and "the second Alaskan 'instant harbor' remained, like Chariot, forever an imagined geography."[13]

By October 1970, all the corporate leaseholders united behind the Trans-Alaska Pipeline System (TAPS) proposal. TAPS would preclude the need for international pipeline right-of-way agreements and offer a variety of transit options. Tankers could travel from Valdez to refineries of the continental West Coast or energy-starved Japan, which received more than 80 percent of its oil from the Middle East. Or tankers could offload their hydrocarboniferous cargo at a proposed pipeline terminal in Central America for reshipping to Gulf Coast and Caribbean refineries. Finally, because the Panama Canal could not accommodate ships larger than 65,000 deadweight tonnage (dwt), supertankers of 70,000 to 189,000 dwt could convey North Slope oil eastward by traveling around Cape Horn or transferring their cargo to smaller ships transiting the Panama waterway, a process known as lightering.[14] Assuming the TAPS plan was a foregone conclusion, the oil companies began ordering pipe from Japanese steelmakers without the required federal permits.[15]

They got a rude awakening when environmental groups sued to halt construction, citing their authority to do so under the National Environmental Policy Act. Congress had passed the law in December 1969, and many of its supporters probably deemed it a mere feel-good nod to the rising environment movement.[16] Even the executive director of the Sierra Club, who had provided the lead testimony before the Senate Interior Committee, later admitted that he "did not foresee the importance of its requirement that agencies document the impact of their proposals on the environment and inform the public of their findings."[17] But environmental groups soon grasped that NEPA's "action-forcing mechanism," the environmental impact statement (EIS), gave them powerful leverage in the courts.

Barely three months after President Nixon signed NEPA on January 1, 1970, the Washington-based Center for Law and Social Policy filed a lawsuit on behalf of three organizations—the Wilderness Society, Friends of the Earth, and the Environmental Defense Fund—to force the Interior Department to conduct a more detailed study of TAPS's environmental repercussions as well as an assessment of alternate routes. The original TAPS report, issued on March 5, 1970, had sparked the litigation because it concluded, based on only eight pages of evidence, that the pipeline would have no major effects.[18]

In what became the first major NEPA-inspired lawsuit, TAPS opponents focused on the threats that the hot-oil pipeline posed to Native American land rights, permafrost, and caribou, as well as the danger of oil spills occurring along the sea route from Valdez and the pros and cons of alternative Canadian routes to the Midwest. In response, the Interior Department issued a new 246-page draft EIS in January 1971, which many citizens outside the scientific and academic communities denounced at public hearings in Washington and Anchorage for failing to take the wide range of potential environmental consequences seriously.[19]

Some environmentally inclined stakeholders invoked other possible unintended consequences. In particular, Minnesota's Democratic senator Walter Mondale, who later became Carter's vice president, led the fight for a Trans-Canadian Pipeline on environmental and economic grounds. Mondale predicted that if TAPS supporters prevailed, a large surplus of oil would flood the West Coast's refineries. He also alleged that the Alaskan oil consortium sought to use the planned Valdez TAPS terminal not to benefit U.S. consumers but rather to export up to a quarter of the precious resource to Japan.[20]

The Interior Department's *Final Environmental Impact Statement*, released in March 1972, covered six volumes. It predicted that the pipeline infrastructural system, including the haul road, oil field, and tanker ballast treatment facility, would have a variety of unavoidable effects during construction, operation, and maintenance. Adverse environmental effects such as permafrost thawing, oil spills, and loss of wildlife habitat would likely occur, but so would socioeconomic changes "which many would classify as beneficial."[21]

Unsatisfied, the plaintiffs used the forty-five-day comment period to gather further evidence and prepare their next round of challenges. Congress initiated new hearings based on a court decision that the report did not meet the requirements of another relevant law, the Mineral Leasing Act. But by July 1973, the pipeline's friends in Congress had had enough. Alaska Senator Mike Gravel introduced an amendment that exempted the project from further environmental review, culminating in a dramatic tie-breaking vote by Vice President Spiro Agnew in July.[22] The path was cleared for the pipeline, and Congress had checked its own initiative to subject megaproject planning on federal lands to greater public oversight.

The TAPS debate held important lessons for environmentalists. Despite their deep disappointment, some took solace in having forced the designers to implement important changes, such as building parts of the forty-eight-inch-wide pipeline above tundra areas vulnerable to melting and around caribou points of passage. They had also forced the oil companies and Interior Department to pay a high price for not proactively complying with NEPA. Equipment sat exposed to the intense Arctic elements for three winters, and government officials received bad press. Advocacy groups also learned about the value of employing lawyers and lobbyists to work the halls of power in Washington, developments that undermined the power of local grassroots approaches but also reaped judicial successes and demonstrated NEPA's status as "a great equalizer in the hands of skilled litigants."[23]

After October 1973, the OPEC embargo made it much harder for environmentalists to mount effective protests against fossil fuel infrastructure projects. OPEC retaliated against the United States for supporting Israel during the 1973 Arab-Israeli War by cutting off the flow of oil, sparking a crippling recession. Environmentalist action turned to conservation and the development of renewable energy sources and more efficient cars and mass transit. In the meantime, TAPS construction proceeded with little attention paid to ensuring that the transportation network that would be needed to distribute North Slope petroleum would meet the same high environmental standards as the pipeline itself.

Reduced U.S. consumption of fossil fuels following the embargo had its own unintended consequences. By the time the monumental eight-hundred-mile pipeline was completed, and the first barrels started flowing in the summer of 1977, federal officials faced an embarrassing dilemma. The predicted West Coast glut had come to fruition, mainly because most of the area's refineries could not handle the North Slope's high-sulfur crude.[24] However, the oil companies' goal of selling the surplus to Japan was now impossible because Congress had banned oil exports as part of the Trans-Alaska Pipeline Authorization Act, any reversal of which would look like an unpatriotic corporate giveaway.[25] As an administration official explained President Carter's quandary, "How can he approve exporting oil to Japan from a pipeline built for national security reasons and still convince the American people that there is an energy crisis?"[26]

Another set of environmental and political issues compounded the problem of transporting oil eastward. Following the Santa Barbara oil rig blowout of 1969, California and Washington had enacted stringent water and air pollution laws that stymied initiatives to build new transcontinental pipelines.[27] Of the four oil infrastructure projects on the table in 1977, the most publicized environmental conflict involved the Standard Oil Company of Ohio (Sohio), controller of 50 percent of the North Slope reserves. Seeking to end the costly work of lightering through the Panama Canal, Sohio proposed to build a new terminal at Long Beach, California, to access an existing natural gas pipeline to transport five hundred thousand barrels of crude oil per day to Midland, Texas, the western terminus of a crude pipeline network stretching to the Northeast.[28] Yet despite the support of the Carter administration, the California Air Resources Board and California Energy Commission objected that the terminal would increase air pollution in the smoggy Los Angeles basin and that repurposing the Midland pipeline for oil would jeopardize the state's natural gas supplies.[29]

The embargo had another shocking effect—for the first time since World War II, the number of ships transiting the Panama Canal fell, from 15,500 in 1970 to 13,500 in 1976. Shippers of petroleum and petroleum products had been a major source of traffic and income for the canal, so much so that the Anderson Commission had surveyed the twenty major U.S. oil companies regarding the nuclear sea-level canal idea in 1968.[30] While the Panama Canal Company preferred to keep tolls low so as to stimulate maritime trade (a custom long deplored by Panama, which received a cut of the proceeds), the need for revenue led it to raise tolls for the first time in its sixty-year history, in 1974 and again in 1976.[31] Shippers complained about the increases, which totaled about 50 percent.[32]

The OPEC ban reduced the number of oil-carrying ships through the Panama Canal, as did the closure of the Suez Canal following the 1967 Six-Day War. The eight-year-long shutdown accelerated the trend toward larger tankers, which could compensate for the longer trips around the Cape of Good Hope by carrying heavier cargoes. Likewise, these superships made it more economical to bypass the Panama Canal in favor of the lengthy South American Cape Horn route. For example, a 100,000-ton ship could load coal at Norfolk, Virginia, iron ore in Brazil, and oil in Nigeria, and then deliver the cargo to Japan by traveling around the southern edge of Africa. Although such a trip took thirty-eight days, it was cheaper than sending a fully loaded 50,000-ton ship for the twenty-five-day voyage to the Far East via Panama. Due to the rapid expansion of supertankers, by 1976 over 1,300 of the world's 22,500 merchant vessels were too wide for the Panama Canal's locks, and another 1,700 ships exceeded the draft limits when loaded to capacity.[33] How should Panama and the United States respond?

The Seaway's Post-PNE Resurrection

The negotiators of the 1964–67 program to replace the Hay–Bunau-Varilla Treaty had devoted one of the three draft treaties to a future sea-level canal. But by the time the two nations resumed serious talks in 1973–75, the futuristic waterway had been all but forgotten. The Anderson Commission's 1970 recommendation against nuclear excavation, the 1971 U.S. National Security Council memo addressing the sea-level canal's ecological issues and lack of military and foreign policy benefits, and the shipping industry's disinterest in any solution requiring high tolls had sapped U.S. enthusiasm for what had once appeared an ideal technopolitical solution. Moreover, a strong sense on both sides that any new canal—however unlikely—would have to be sited in Panama dissipated the U.S. leverage that had prevailed in the previous treaty talks.[34]

Even so, the diplomats retained language allowing for eventual cooperation on a sea-level waterway. As U.S. co-negotiator Sol Linowitz later explained, when the treaty negotiations had resumed under President Nixon, "Panama had agreed to give the United States the right to build a sea-level canal, and our draft of the principles had locked it up in iron, as lawyers do when an issue is really moot and there has been no dispute about it."[35] The principles, known as the 1974 Kissinger-Tack Agreement after Secretary of State Henry Kissinger and Panamanian Minister of Foreign Affairs Juan Antonio Tack, provided guidelines for nullifying the 1903 treaty, including one that future efforts to increase the canal's capacity would be bilateral. Subsequently, in 1975 President Ford directed his negotiators to seek the longest possible period for a U.S. option to increase canal capacity, either by adding a third lane of locks or building a sea-level canal.[36] It was not that Ford sought to build a new waterway anytime soon, but rather that he wanted to prevent any other nation—particularly Japan, the canal's second-largest user—from doing so, since technological and economic changes might eventually render such a project viable and useful for the United States.[37]

Ford suspended the negotiations as the 1976 presidential campaign heated up, during which his Republican primary opponent, former California governor

Ronald Reagan, discovered that the issue of maintaining U.S. control of the Panama Canal made for a powerful talking point among conservative audiences.[38] However, Democrat Jimmy Carter won the election, with strong support from environmentalists. He immediately made the Panama Canal Treaties a centerpiece of his foreign policy agenda, retaining Ford's negotiator Ellsworth Bunker (a former ambassador to Vietnam) and appointing Linowitz (a former ambassador to the Organization of American States) as co-negotiator, for a six-month term to end on August 10, 1977.[39]

Carter's action upset Representative Dan Flood, then in his twenty-second year in the House. Treaty reform, Flood warned in a private letter, "could well be your 'Bay of Pigs' and prevent your renomination or re-election," an ominous reference to President Kennedy's disastrous failure to overthrow Fidel Castro's government in Cuba in 1961. Flood also tried to appeal to Carter's environmentalism by invoking the biologists whom he had previously invited to testify: "The old idea of a sea level canal is irrelevant and strongly opposed by major conservation organizations, as well as engineers, because of the danger of infesting the Atlantic with the poisonous Pacific sea snake and the crown of thorns starfish as well as the other factors."[40]

Flood had good reason to presume Carter would find the biological rationale compelling. During the 1976 campaign, Carter's team had cultivated the environmental vote by hailing his record as governor of Georgia. After being persuaded by environmental and conservation groups to reverse his support for an imminent Army Corps of Engineers dam on the Flint River, Carter incurred the wrath of many Georgians by vetoing the project. But it raised his national profile, and solidified his resoluteness as chief executive to eliminate wasteful pork barrel water projects—a goal that combined his interests in environmental quality and fiscal conservatism.[41]

Flood, a conservative Democrat, had little interest in environmental issues—except for when it came to his passion project of preserving the Panama Canal Zone. He and other opponents of treaty reform had been mobilizing the ecological arguments against the sea-level canal proposal for years.[42] At a 1973 hearing, his Republican colleague, South Carolina senator Strom Thurmond, also used the concerns of the Smithsonian scientists to advance his own reasons for expanding the existing waterway: "The American housewife is already feeling the effects of the disappearance of Peruvian anchovies, apparently from overfishing, which were a major source of cheap fishmeal for chicken feed. Opening up the isthmus to a sea-level passage could well be opening up a Pandora's box for the world's food supply." By contrast, Thurmond argued (not incorrectly), "major environmental groups look upon the Terminal Lake–Third Locks plan as a positive step in averting ecological disaster."[43] But because conservative treaty opponents did not otherwise support environmental causes, the scientific community saw through their selective use of their data.[44]

Unlike in the 1960s, by the mid-1970s the U.S. environmentalist community had an institutional base that several of the largest organizations used to assert a voice in the growing debates over modernizing the canal and U.S.-Panama relations. Friends of the Earth (FOE) played a prominent role, beginning in 1973 when the sea-level canal idea attracted renewed attention at congressional hearings on the efforts of Flood, Thurmond, and others to allocate $850 million for a third lane of locks for the original waterway.[45] FOE was a young organization, having been founded in 1969 by the former executive director of the Sierra Club David Brower. Taking strong stands against TAPS, chemical warfare in Vietnam, nuclear energy, and other technologies of the postwar era, the group's membership grew to twenty-seven thousand within four years.[46] Its legislative director, George Alderson, had become such a congressional fixture that a member of the House Interior committee joked about calling him by his first name. Alderson had also helped organize the successful environmentalist coalition against the federal supersonic transport program, and later authored a guide to citizen lobbying.[47]

During both the Ford and Carter administrations, as a leading employee of FOE and then of the Wilderness Society, Alderson endeavored to mobilize the insights of the 1960s-era marine biologists. As a fellow environmentalist put it in 1977, "George Alderson has done the nation and the world great service by paying more attention to the Sealevel Canal [sic] than anyone else in the conservation-environmental establishment."[48] The former biology major, who had begun his career in the late 1960s at the Sierra Club and then followed Brower to FOE, became alarmed when the president of the Panama Canal Company testified before Congress in 1973 in mild support of building a sea-level channel at an unspecified future point.[49] While shipping industry representatives objected that such a waterway would lead to increased tolls, Alderson sought to elevate ecological matters to the level of economic ones. In letters to the House Panama Canal subcommittee chairs requesting the opportunity to testify at future hearings, Alderson asserted that expanding the original waterway would serve commercial maritime needs "without allowing disruption of the marine ecosystems in the adjacent waters," and noted that no environmental impact statements for modernizing the Panama Canal had yet been filed.[50] He included a recent article from *Defenders of Wildlife News* by John Briggs, the University of South

Florida zoogeographer who had described the seaway in 1969 as a potential biological catastrophe. Briggs had since participated in an international conference devoted to the biological effects of interoceanic canals, held in Monte Carlo in 1972, and remained convinced that a seaway could cause "a huge and irrevocable loss of perhaps thousands of species native to the Eastern Pacific."[51]

Alderson had spearheaded an earlier censure of the sea-level canal in April 1975, when he helped organize ten environmental groups to telegram President Ford that "it would be premature and reckless to enter a new treaty with Panama that would authorize or permit construction of a sea-level canal."[52] Two years later, soon after President Carter took office, Alderson began lobbying the administration to integrate environmental concerns into the treaty negotiations and to reject the sea-level canal project once and for all.

Alderson tried to get the White House's attention by writing to the Domestic Policy Staff's environmental expert, Katherine Fletcher. She had participated in Earth Day 1970 and, after completing her undergraduate biology degree, worked in Washington lobbying for environmental legislation and in Colorado opposing oil-shale development there.[53] In March 1977, she was consumed with the congressional fallout of the water projects hit list, among other issues.[54] But the seaway, Alderson emphasized, would be far worse than any of the dams in question: "I'm not sure whether it's in your bailiwick, but the government has been edging into a $3-billion water project that could be a bigger fiasco than any we've seen yet. It's the proposed Sea-level Panama Canal, which is an almost unnoticed topic of the Panama treaty negotiations."[55] Alderson asked her to consult with three agencies regarding the seaway's environmental and economic effects: the Council on Environmental Quality (CEQ), the executive agency responsible for reviewing environmental impact statements; the Office of Management and Budget; and the National Oceanic and Atmospheric Administration, the employer of a scientist who had coauthored an important article about the ability of canals to facilitate nonnative species exchange.[56] Fletcher responded two weeks later, in April 1977, that she was forwarding his letter to CEQ and Office of Management and Budget and would like to learn more about this "very interesting situation," but otherwise remained noncommittal.[57]

The following month, Carter received a persuasive letter from a pro-sea-level canal senator, Mike Gravel. The Alaska Democrat had gained a national following among liberals for entering the Pentagon Papers into the *Congressional Record* in 1971, but angered environmentalists for sponsoring the 1973 amendment to exempt the Trans-Alaska Pipeline Authorization Act from further environmental review. Gravel had learned about the sea-level canal just recently, during a March 1977 fact-finding trip to Panama. He "picked up the cause," having become convinced that it offered the best solution to the problem of distributing Alaskan oil to the energy-hungry East and Gulf Coasts.[58]

In his May 5 letter to the president, the senator advised delaying the treaties' ratification because the question of who possessed sovereignty over the Canal Zone remained a "burning emotional issue in both the United States and Panama." In the meantime, he said the administration should work with Congress to authorize Carter to "take certain unilateral steps to improve the situation with Panama." In particular, the Army Corps of Engineers should update the Anderson Commission's 1970 report. Other recommended actions echoed those of previous administrations, such as transferring some lands and responsibilities for operating the canal to Panama.[59]

A civil liberties framework infused Gravel's ensuing forty-six-page report to the Senate Environment and Public Works Committee, which outlined the political and technological outmodedness of the existing waterway. He focused on the need to rectify the historical injustice perpetuated by the United States against Panama and argued that the U.S. did not own the Canal Zone, a "colonial-socialistic enclave" incommensurate with American values. Gravel rejected the argument that Panamanian control would lead to the waterway's ruin. He also presented data he had commissioned demonstrating the canal's diminishing value due to its inability to accommodate the new class of post-Panamax supertankers.[60]

Having outlined the case against retaining U.S. control of the existing canal, Gravel presented five reasons for developing a new interoceanic waterway at sea level, any one of which in his view justified the project. First, the need to transport the oil and natural gas resources of his own state eastward would enhance the economic viability of a wide sea-level canal. Such a waterway, which he proposed to be fully owned by Panama, would resolve the thirteen-year-old stalemate in treaty negotiations, reduce U.S. dependence on foreign energy sources, preclude massive U.S. investments in new east-west oil pipeline systems, and strengthen U.S. defense capacities by allowing the navy's thirteen post-Panamax aircraft carriers to transit in the event of a national security crisis.[61]

Although he addressed the report to the Senate Committee on Environment and Public Works, and despite President Carter's well-known environmental concerns, Gravel made no mention of the furor over the biological questions raised by the 1960s-era sea-level canal debate. Nor did he ground his concerns about energy security and infrastructure needs in the language of environmental issues. Alderson was livid, having learned about Gravel's concurrent efforts to secure the support of the Army Corps of Engineers and Department of Transportation for a three-year, $7 million reanalysis of the seaway. As Alderson notified Fletcher in mid-May, "His staff was not well informed on the environmental problems involved, but explained that all he wants is a study—the old refrain."[62] One of Alderson's colleagues also pressured Fletcher to persuade her bosses "to oppose this turkey," which Gravel had probably forgotten had begun "as an AEC pipedream."[63] Responded Fletcher, "You can be assured that the Administration would not be lightly led into the support of this multi-billion dollar project," a letter that would later be used by FOE in a congressional hearing.[64]

Gravel's lobbying paid off when he scored a twenty-minute meeting with the president in the Oval Office on July 13, 1977.[65] In preparation, the State Department urged Carter to affirm that they could win ratification via a massive public relations campaign, and that waiting any longer to take action on the treaty would risk violence with Panama and "offer domestic opponents opportunity to torpedo it." The briefing paper's final point reminded Carter to tell Gravel that the new treaty would provide an option for the U.S. to build a sea-level canal, but otherwise the document's background section included no information about the proposal's checkered history.[66]

That seemed to be the end of it, but eight days later, on July 21, 1977, Carter brought up the sea-level canal idea in an unexpected venue. At a town hall in Yazoo City, Mississippi, a citizen implied that he opposed granting Panama control of the canal and relinquishing the Canal Zone because the area would provide vital military assistance in the event of a third world war. Carter explained his rationale for new diplomatic arrangements between the two countries, and then went a step further by stating, "My guess is, that before many more years go by, we might well need a new canal at sea level, that can handle very large ships." While President Lyndon Johnson's administration had studied the multibillion-dollar project, Carter explained, the need to transport Alaskan oil and gas in huge tankers to the Gulf and East Coasts had since then intensified the need for a wider waterway.[67] The following day, after attending an energy conference in New Orleans and helicoptering to an offshore oil rig, Carter elaborated that the Alaskan oil situation had transformed the notion of what kinds of infrastructure were economically "shocking and unreasonable." The $8 billion spent by private industry on TAPS, and the $12 billion projected for an Alaskan natural gas pipeline, had put the Anderson Commission's estimate in perspective.[68] The $2.88 billion cost of the Route 10 seaway had seemed beyond the pale in 1970, but now, even when adjusted for inflation at $5.29 billion, it seemed much less objectionable.[69]

The president's invocation of the futuristic seaway surprised everyone involved, even himself. As the *New York Times* reported the day after the Yazoo City event, "Administration officials said today it was possible an option for a sea-level waterway might be agreed to as part of the negotiation for a new canal treaty, but they were obviously startled that President Carter seemed to treat it as a matter under serious consideration. Senator Mike Gravel, Democrat of Alaska, had revived the idea recently."[70] Journalists stopped short of sharing Carter's admission to having devoted little attention to the idea until then. At a question and answer session with members of the press in New Orleans on July 22, after one reporter asked for more details regarding his rationale for supporting a new oil-conveying seaway, Carter responded, "I told you at least as much as I know," prompting laughter from the crowd. He then made the point that a new canal would not be exorbitant compared to other alternatives and the $8 billion price tag for TAPS, concluding, "I've not gone into the question in any depth and I'm not prepared to answer any further."[71]

Staff members of the White House Office of Science and Technology Policy scrambled to meet the request of their boss, geophysicist Frank Press, to prepare a short memo on the seaway's ecological effects. As one wrote, "Both Frank and I remember that it was suggested that there could be some rather serious ecological consequences—sea snakes on the Pacific or the Atlantic side (one or the other) making their way to the other ocean."[72] The ensuing document to the president began with the words, "Your recent statements on a sea level canal in Central America will revive discussion of the potential environmental effects of such an endeavor," and summarized the 1970 assessments of the Anderson Commission and Mayr Committee. Press concluded with an offer to initiate an update by the National Academy of Sciences, which Carter approved.[73] During the late summer and fall of 1977, a multitude of agencies and organizations rushed to contribute to the revived discussion.

Article XII

Carter's sudden espousal of the sea-level canal annoyed the treaty negotiators, who considered it settled and subordinate to resolving differences over the U.S. military bases, canal annuities to Panama, and many other issues. It now became the major obstacle to reaching a final agreement.[74] Bunker and Linowitz warned Secretary of State Cyrus Vance that revising the sea-level canal provision would harm the negotiations by signaling a stronger U.S. intention to build it than the

Panamanians had been led to believe, and by diminishing potential congressional support for the new treaty.[75]

On the other hand, argued the assistant secretary of state for congressional relations, the prospect of future access to a sea-level canal might induce some reluctant senators to vote for the treaty: "People may feel more comfortable if they see the possibility that we and the Panamanians may be building toward a promising joint enterprise in the future rather than simply disengaging from an unsatisfactory past relationship."[76] Following a meeting with several senators a week later, Vice President Mondale informed the White House chief of staff that "there was considerable sentiment for a new sea level canal" and thus "we should consider sharpening and strengthening the language in the current draft under negotiation which would give the U.S. right of first refusal for a new canal."[77] An unnamed administration official reiterated this idea by telling a journalist, "Politically, a sea-level canal in the treaty makes it a little easier to sell to Congress."[78] As the protreaty Senator Fritz Hollings explained to his South Carolina constituents, due to rising construction costs and the need for larger canal capacity, "With hindsight now we realize that rather than working for thirteen years to renegotiate the old treaty, we should have insisted on a new sea-level canal. This would have been wide enough for all our warships as well as the largest oil tankers. Then the sovereignty, sabotage and other problems would have been moot."[79] Of course, the 1964–67 negotiations did give the U.S. the right to build such a waterway. Despite his historical amnesia, Hollings exemplified the venerable view that complex political problems could be resolved via a technological approach.

But shoehorning the sea-level canal provision into the almost-complete treaty was complicated. Bunker and Linowitz tried to obtain a U.S. veto over canal construction by a third country in Panama until the treaty's expiration on the last day of 1999. In response, at an August 5 conference in Bogota to finalize the Panamanian terms, Torrijos and the presidents of Colombia, Venezuela, Mexico, Costa Rica, and Jamaica called for the United States to reciprocate by agreeing not to negotiate with other Central American nations to build a new interoceanic route during the life of the treaty. The Latin American presidents appealed to Carter to accept the condition; otherwise, Panama would walk away from the table.[80]

The Panamanian negotiators later charged that U.S. insistence on the right to exclude other nations from excavating a sea-level channel through Panama jeopardized the treaty talks, an allegation that U.S. officials downplayed. Whether it constituted a full-blown "crisis" or routine "horse trading," the issue did incite intense discussions at the eleventh hour.[81] After poring over the Anderson Commission report, Carter's advisors privately noted, "There is little likelihood that if we chose to build a sea-level canal it would be anywhere else but Panama. Although it might be argued that the option to build a canal in a third country gives us added leverage over Panama, any hint of using such leverage would provoke such an adverse reaction in Latin America that, in effect, we couldn't use it."[82] Extracting the kinds of concessions that had appeared possible during the previous decade was now out of the question.[83]

For others, however, the forfeiture of U.S. rights to negotiate with rival Central American nations for a sea-level route made no sense. In his case for the canal treaties, for instance, Hollings ended his spiel for the sea-level waterway by asserting, "What is unexplainable is the provision that forbids us to negotiate a new canal anywhere but Panama."[84] South Dakota representative Larry Pressler cited the 1970 Canal Study Commission report to argue for the Colombia route due to its lower cost and ability to "break up Panama's monopoly on interoceanic transit, which would tend to reduce the potential for economic blackmail."[85] The *Chicago Tribune*, the newspaper that had published leaked drafts of the 1967 treaties, described the provision that obligated the U.S. not to build a canal through another country as "a real puzzler" and an indication that "Panama must have some reason to think we might not be happy with the way the [present] canal is run. . . . Like the wealthy lady whose fiancé refused to marry her unless she gave him power over her investments, we find this treaty provision distinctly disconcerting."[86]

By rejecting the engineering and political assessments that Panama offered the best place for a seaway, such arguments overlooked the possibility of another nation (Japan especially) cutting its own deal with Panama, a prospect that concerned Carter. As he later explained in one of many presentations designed to sell the treaties to skeptical audiences, "This is a clear benefit to us, for it ensures that, say, ten or twenty years from now, no unfriendly but wealthy power will be able to purchase the right to build a sea-level canal, to bypass the existing canal, perhaps leaving that other nation in control of the only usable waterway across the isthmus."[87]

The final compromise over the sea-level canal provision denied the United States an exclusive right to build a waterway anywhere but in Panama while granting the U.S. veto power over construction of such a route in Panama by any other nation before the end of the century. In the sardonic words of Ambassador Linowitz, who conveyed his exasperation over the "implausible prospects of the sea-level canal" in his 1985 memoir, "in effect it added up to the same thing."[88] Later in August 1977, after the announcement that the two countries had reached agreement on all the major diplomatic principles, the *New York Times* published an embarrassing account of the frustrations that Carter's Yazoo City comment had caused for both diplomatic teams.[89]

Another awkward, unforeseen consequence of Carter's words played out as dozens of Latin American and U.S. dignitaries assembled in Washington, D.C. to celebrate the signing of the Torrijos-Carter Treaties. The day before the ceremony, on September 6, 1977, eleven organizations spanning the spectrum of environmentalist advocacy telegrammed the White House with an urgent plea to reject Article XII of the Panama Canal Treaty. The majority of the telegram participants were decades old and well known for their advocacy of protecting natural resources and habitat: the National Parks and Conservation Association, Izaak Walton League, Wilderness Society, Defenders of Wildlife, and Federation of Western Outdoor Clubs. A second subset of the signing groups embodied the newer focus on pollution prevention: Environmental Action Inc., the Environmental Policy Center, and FOE, which probably led the telegram effort. The third subgroup consisted of organizations devoted to animal welfare: the World Wildlife Fund, Fund for Animals Inc., and International Society for the Protection of Animals.

The telegram conveyed information both old and new. It quoted from the 1970 report by the National Academy's Committee on Ecological Research for the Interoceanic Canal (CERIC) and reminded the president of the 1975 telegram opposing any treaty option involving a sea-level waterway. The coalition objected that the State Department had only released a draft EIS of the Panama Canal Treaty days earlier and that the academy had not yet issued its updated report for the White House. Due to the lack of information needed to address such a serious environmental issue, they argued the government should reject the option to develop that which represented "not only an economic loss but a likely environmental disaster as well."[90]

A parallel media outreach initiative resulted in a nationally syndicated editorial subtitled "A Passage to Ecological Disaster." The op-ed summarized the major findings of the 1960s-era debate, called out Senator Gravel for ignoring the issues raised by CERIC, and observed that "President Carter's expressed enthusiasm for the eventual construction of a sea-level Panama Canal doesn't sound like the Jimmy Carter that environmentalists have come to know, and more often than not, admire."[91]

Indeed, less than four months had passed since Carter had delivered a major environmental message to Congress. The thirty-six-page document outlined a

FIGURE 7.1. Panamanian general Omar Torrijos and U.S. president Jimmy Carter at the signing of the Panama Canal Treaties in Washington, D.C., September 7, 1977. Marion S. Trikosko, photographer. Prints and Photographs Division, Library of Congress, LC-DIG-ppmsca-09785.

wide range of initiatives pertaining to pollution control, human health, energy, the urban environment, and population growth. It even included a plan to restrict "the importation of exotic species." As he explained, "In the past 150 years, hundreds of foreign wildlife species, both plant and animal, have been introduced into the natural ecosystems of the United States," many of them "highly detrimental to public health, agriculture, and native wildlife." Consequently, on that date, May 23, 1977, Carter issued an executive order to prohibit the establishment of exotic organisms on federal lands and waters, and directed the secretaries of agriculture and the interior to develop legislation to address the problem.[92]

Although Article XII seemed to contradict key environmental goals of the Carter administration, the environmentalists' last-minute telegram did not stop the signing ceremony for the treaties. The event took place on September 7 in the grand Hall of the Americas at the Washington headquarters of the Organization of American States (fig. 7.1). Eighteen Latin American heads of state attended, as well as a bipartisan group of U.S. dignitaries including President Ford, President Johnson's widow Lady Bird Johnson, and former secretaries of excavation with unexpected diplomatic and economic issues, required much more than the three years originally allocated by Congress.

Key allies enabled the commission to gain one last extension, until December 1, 1970. Except for Flood and a few others, even some of the stalwart congressional defenders of the canal status quo, especially Representative Leonor Sullivan of the pivotal House Merchant Marine and Fisheries Committee, worked to ensure the completion of the sea-level canal studies.[98] In the process of plotting out their final two years, the members of the Anderson Commission began shifting focus to another route that ruled out PNEs. Yet even as they became more realistic about the political infeasibility of nuclear methods, other important issues in the air blindsided them.

CHAPTER 5

Assessing Mankind's Most Gigantic Biological Experiment

THE YEARS OF 1968 through 1970 were as momentous for the Anderson Commission as for the rest of the United States. The increasingly uncertain political viability of the nuclear cratering experimental program consumed their attention, along with two other major issues: a geological realization and a marine biological awakening. The geological and marine biological dimensions of the sea-level canal endeavor prompted significant epistemological and diplomatic challenges and opportunities, and they show how the isthmian environment itself—both the land and surrounding seas—helped shape political perceptions of the new waterway's feasibility.

The commission sponsored several data-collecting expeditions along the two proposed nuclear routes. Because scientists knew so little about the Darién rainforests and marshlands, much of the work entailed basic environmental research—quantitative studies that Humboldt would have approved.[1] At the peak of the endeavor, over eight hundred U.S. and Panamanian researchers toiled in the tropical heat, measuring and recording variables pertaining to topography, geology, hydrology, hydrography, forest ecology, and anthropogenic food chains. Corps of Engineers personnel also conducted experiments pertaining to ground and air blast activity. The subcontracted studies generated numerous peer-reviewed journal articles and volumes of gray literature.[2] They also employed many young researchers who built prominent careers in ecology, ethnobotany, anthropology, and civil engineering.[3]

The work constituted routine, normal science—with one exception.[4] Near the end of the surveys in Panama, workers confirmed that the saturated clay shale soils along a twenty-mile stretch of the otherwise-ideal Route 17 would not hold up to underground thermonuclear blasts. While not wholly unexpected, the finding had revolutionary implications both for the proponents of nuclear construction and for the U.S. treaty negotiators. How the commissioners dealt with the clay shales "bad actor" sheds light on the secretive, technocratic aspects of their approach to environmental impact assessment.[5]

colonization. He also questioned the scanty empirical evidence for ballast-mediated carriage, arguing that Menzies used intertidal organisms preadapted to survival in fresh water and that Chesher overlooked the toxic anticorrosion coatings lining ballast tanks. He did concede that modern tankers with stainless steel tanks—most of which were too large to use the existing canal—might enable plankton to survive passage from ocean to ocean. "The actual role of ballast transport through the present Canal is a subject that *could* be properly evaluated, and a thorough study should remove this area from speculation," he argued, as did other sea-level canal authors.[100]

Of course, such research could be conducted only if funding came through, and to that end, the CSC and Smithsonian brokered a strategy to try to resolve the problem that had occupied Smithsonian naturalists for the past half decade. The commission scrounged up enough funds to elicit the National Academy of Sciences to lend its authority to the quest for external federal support of a comprehensive tropical marine biological inventory.

Conclusion

The CSC conveyed its fifth annual report to the new president, Richard M. Nixon, in the summer of 1969. The report acknowledged the Route 17 clay shale problem, as well as the expanded efforts to address marine biotic exchange. However, like his predecessor Nixon neglected these points in his public message when he forwarded the report to Congress, focusing instead on the good news: the engineering feasibility team had completed data collection efforts and closed down all field operations; the diplomatic, economic, and military subgroups were wrapping up their evaluations; and the Plowshare scientists had conducted the third 1968 cratering experiment, Project Schooner, at the Nevada Test Site. While conceding that all six planned shots would not be completed prior to the December 1970 deadline, the commission still expected to render its conclusion by then on the feasibility of nuclear explosives for canal excavation.[101]

Neither the CSC report nor the president, however, mentioned another setback, the uncontrolled release of radioactivity by the 35-kiloton Schooner explosion. Five days after creating a crater of 725 feet wide and 250 feet deep on December 8, 1968, the experiment caused radiation levels at sampling stations as far away as Ontario and Quebec, Canada, to rise to ten to twenty times above normal background levels.[102] The fifth annual report also glossed over Panama's revolutionary events of October 1968—the military coup staged by Omar Torrijos and other officers of the Guardia Nacional. Soon after taking control of the government, Torrijos established his own Office of Interoceanic Canal Studies and ordered an investigation into a Panamanian-built seaway using nonnuclear methods along the Route 10 site west of the existing canal. While hinting at the recent political changes, the international relations section of the CSC document praised the U.S., Panamanian, and Colombian officials whose cooperation made the field surveys possible. "A large quantity of environmental information has been acquired in areas of the isthmus that previously had been little explored," valuable data that would soon be made public.[103]

Yet from a high-modernist standpoint, new scientific knowledge of the mythic Darién landscape failed to counterbalance the increasingly bad news about PNEs. Reported a *Panama American* journalist in the spring of 1969, "an atomic engineers' dream . . . is fading into a ditch diggers' pick and shovel nightmare."[104] That history could move backward, from the space age to the olden days, seemed incredible to the friends of Plowshare. So did the insistence of biologists that the canal engineering feasibility studies address issues that had nothing to do with radioactivity, just so they could resuscitate the old-fashioned study of natural history.

Today, fifty years on, it is clear that those demanding realistic, comprehensive assessments of the megaproject's environmental effects were as forward-thinking as the other sea-level canal stakeholders. The Plowshare physicists and allied engineers sought to advance the unproven field of nuclear excavation, and U.S. officials aimed to update the isthmian transportation system and relations with Panama. Likewise, the marine evolutionary and ichthyological scientists sought to revitalize relevant naturalist disciplines to analyze the enormous ecological changes a wide channel at the level of the seas would induce. The sea-level canal proposal served as a modernization strategy for a wide range of stakeholders, and it facilitated many kinds of work with far-reaching political and intellectual effects.

state William Rogers and Henry Kissinger. Near the end of his brief remarks, President Carter stated, "In the spirit of reciprocity suggested by the leaders at the Bogota summit, the United States and Panama have agreed that any future sea-level canal will be built in Panama and with the cooperation of the United States. In this manner, the best interests of both our nations are linked and preserved into the future."[93] Otherwise neither he nor Commander Omar Torrijos mentioned the environmental implications of the two treaties, an unsurprising omission given their preeminent focus on military and economic security.

But as the diplomatic process gave way to domestic politicking for the ratification of the accords, Carter could not ignore the environmentalists' antiseaway publicity indefinitely. Although he had embraced Senator Gravel's vision of the sea-level canal as a utilitarian solution to serious problems regarding the U.S. energy supply, the proposal, as it had in the 1960s, raised difficult questions about the relationship between oceanic health and maritime infrastructure. The Senate ratification campaign would soon force him to realize that federal officials could no longer dismiss such esoteric questions as easily as in the pre-NEPA years of the previous decade.

Conclusion

Although the emergence of nuclear excavation technology had revived the dormant dream of an isthmian sea-level canal in the late 1950s, the demise of PNEs in the early 1970s did not lay it to rest. In fact, the seaway idea outlived by several years the postwar enthusiasm for using buried thermonuclear bombs to overcome the economic and technical burdens of monumental engineering projects. Project Plowshare funding finally ended in 1977, seven years after the Anderson Commission concluded it could not recommend atomic ditch-digging. While conventional construction did pose much higher costs than those originally calculated for peaceful nuclear explosives, by the mid-1970s the economic and political calculus had changed. The $8 billion Trans-Alaska Pipeline created a new benchmark for public and private stakeholders about what constituted reasonable expenditures for critical fossil fuel facilities and infrastructure.

The bioenvironmental context had changed as well. The Alaskan oil pipeline, in concert with the National Environmental Policy Act of 1969, set new expectations about the proper scope of what the 1965–70 Anderson Commission had called "environmental considerations." No longer could a four-page assessment such as theirs suffice, since NEPA enabled environmentalist groups to sue federal agencies for failing to evaluate the potential environmental impacts of projects planned for public lands or with public funds. The men appointed by President Johnson to investigate the seaway's feasibility would have been amazed by the multivolume EIS issued by the Department of the Interior and Federal Task Force on Alaskan Oil Development in 1972.

Granted, Congress had demonstrated its authority to check the rapid rise of environmentalist influence on federal megaproject planning by exempting TAPS from further such review in the summer of 1973. Perhaps President Carter had this in mind, as well as the shockwaves caused by the subsequent OPEC embargo, when he stunned his advisors by speaking out in favor of a treaty clause regarding U.S.-Panamanian cooperation to develop a second canal at the level of the seas.

Soon after the signing of the Torrijos-Carter Treaties, an anonymous administration official explained the president's surprise announcement at the July 21 town hall as a function of Carter's "desire to establish himself as a statesman of vision and an activist for new, challenging ideas and projects."[94] Despite the obvious public relations framing, the statement rang true in many respects. Even in his May 23 environmental message to Congress, Carter had emphasized the need to develop the outer continental shelf's federal oil and gas reserves "in an orderly manner, reconciling the nation's energy needs with the fullest possible protection of the environment."[95] Carter's techno-enthusiasm had deep roots; before resigning his commission as a naval officer in 1953, he worked under Admiral Hyman Rickover on the early stages of the U.S. nuclear submarine program.[96] His subsequent success in agribusiness also featured early adoption of cutting-edge technologies, from prescription fertilizers to streamlined peanut-shelling equipment.[97]

A reporter described Governor Carter in 1971 as "an enigma and contradiction," an assessment that still applies to significant aspects of his presidency, including his environmental record.[98] The notorious water projects hit-list episode was but one case that confounded his environmentalist constituency. In August 1977, outraged congressmen from both parties presented Carter with an appropriations bill that restored support for all the water projects for the 1978 fiscal year. The president's reluctant signing of the bill frustrated environmentalists, as did later initiatives to address the energy crisis, including one to fast-track synthetic fuel plants, refineries, and facilities.[99] Carter's support for the sea-level canal should be seen within this context of supporting federal environmental regulations and projects, but not when they conflicted with his top priority of economic recovery.[100]

Understanding why Carter endorsed the sea-level canal in spite of its well-publicized ecological risks requires us to consider not only the dire economic conditions of the 1970s but also the still-inchoate intellectual and political context of marine conservation biology. As revealed by his executive order restricting the introduction of foreign plants and animals, Carter possessed a keen appreciation of the potential threats posed by nonnative invasive species. However, the executive order's framing with respect to the federal agencies overseeing agriculture and public lands, and his own farming background, suggests that Carter and his environmental lieutenants viewed nonnative species through a terrestrial rather than marine lens.

The specific insights of the late 1960s-era biologists regarding marine species exchange profoundly influenced environmental organizations opposing the sea-level canal in the mid-1970s, and yet the science of marine invasion biology—and marine conservation biology more broadly—remained underfunded and underdeveloped. Oceanography, fisheries science, and other marine disciplines still focused more on providing data to facilitate the development of ocean resources than on elucidating their biological diversity and vulnerability to human activities.[101] Along these lines, it is possible that Carter's time in the navy during and after World War II—when naval officials, the shipping industry, and scientists alike sought to use the oceans, whether for fish and mineral extraction, nuclear waste storage, or national security purposes—shaped a view for him of the oceans as invulnerable to overexploitation.

That is not to say that Carter, nor others who sought to maximize the productivity of the seas, did not care for the life within. As a sonar officer searching for Soviet subs with state-of-the-art listening equipment, he learned how to identify dolphins, shrimp, and different species of whales and fish based on their "chatter and songs." Later in his life, Carter shared moving details of his seven years in the navy: "Before submarines were equipped with nuclear power and snorkels, they stayed mostly on the surface. Our hours on the bridge, in the conning tower, or in the sonar room allowed each of us to know the ocean and the heavens in a unique way."[102]

Of course, knowing the ocean from the vantage point of a Cold War submarine did not necessarily facilitate an appreciation for the murky details of marine bioinvasions. Carter's naval national security worldview, engineering mindset, trusteeship approach to governing, and commitment to solving the energy crisis by both conserving and producing energy—as well as his need for canal treaty ratification votes from every possible source of support—likely all played important roles in his advocacy of the sea-level canal.

CHAPTER 8

Containing the Panama Canal Treaty's Environmental Fallout

WITHIN THE FIRST NINE MONTHS of assuming office, the Carter administration succeeded, despite enormous obstacles, in crafting an agreement with Panama to replace the 1903 Hay–Bunau-Varilla Treaty and its 1936 and 1955 supplements with two new accords, the Neutrality Treaty and the Panama Canal Treaty. Next on the agenda was another titanic challenge, securing Senate approval of the pacts before the 1978 midterm elections. In what became one of the most contentious foreign policy debates and longest public relations efforts in U.S. history, high-level officials and surrogates met with hundreds of constituent groups and opinion makers, and cut deals with dozens of individual senators, to secure the sixty-seven senate votes needed for ratification.[1] Carter and his team spent considerable political capital to reassure reluctant Americans that the accords did not constitute an abject surrender of U.S. prerogatives. Much less appreciated in this process was the role played by environmental groups, who made it clear after Carter revived the sea-level canal proposal that their support could not be taken for granted.

The Panama Canal Treaty, which entered into force on October 1, 1979, and terminated on December 31, 1999, with the transfer of full control of the waterway to Panama, contained two major environmental provisions. Article XII authorized the two nations to conduct a feasibility study for a sea-level waterway, prohibited Panama from building a new interoceanic canal and the United States from negotiating with other Central American countries to construct one, allowed the U.S. to add a third lane of locks to the existing canal, and banned the U.S. from using nuclear excavation techniques without Panama's express consent. The less contentious Article VI provided for a joint oversight commission to ensure that the two nations implemented the pact "in a manner consistent with the protection of the natural environment of the Republic of Panama."[2]

Alarmed by the revived prospect of the sea-level canal, savvy environmentalists invoked the National Environmental Policy Act of 1969 (NEPA) to pressure Carter officials into an unprecedented action—drafting an environmental impact statement (EIS) for a treaty that primarily addressed issues of sovereignty and economic and military security.[3] NEPA had already transformed statist environmental management by enabling citizens to take legal action against agencies that failed to take the EIS process seriously, as epitomized by the 1970 Trans-Alaska Pipeline proceeding against the Department of the Interior and the 1971 Calvert Cliffs nuclear reactor licensing lawsuit against the Atomic Energy Commission.[4] As for the law's applicability to international conventions, prior to 1977, the only treaties to undergo formal impact assessments by U.S. agencies focused on unequivocal environmental themes such as natural resource use, pollution control, and the endangered species trade.[5] The State Department's completion of an EIS for the Panama Canal Treaty thus marked a milestone in environmental diplomacy.

The report was not a paragon of rigorous analysis, but it provided a new forum for addressing the ecological threats posed by the sea-level canal. A decade after Ira Rubinoff and fellow biologists had tried to mobilize funds for taxonomic and evolutionary studies needed to predict the seaway's effects on marine fauna, they participated once again in government initiatives to assess the megaproject, but now with the support of professional advocacy nongovernmental organizations (NGOs) and the presidential science advisor.

Not only did environmentalist mobilization against Article XII result in the rushed draft EIS and a new National Academy committee report, it also helped set in motion a postratification set of House hearings on whether to update the 1965–70 Anderson Commission report in accordance with NEPA. Although many members of Congress doubted the eventual need for a second canal, the continuing energy crisis made them more willing to hear out Senator Mike Gravel's proposed remedy.

The 1970s-era phase of the sea-level canal story illustrated the expanding international scope of U.S. environmentalist advocacy. Some environmental arguments and rhetoric, however, did not reflect the Panama Canal Treaty's anti-imperialistic spirit. Rather than addressing the environmental impacts of the different management options available to Panama, the draft and final versions of the EIS delineated a dire scenario that presumed Panamanian mismanagement: if the treaty's environmental protection measures failed, then "the forests and associated ecosystem in the Canal Zone could disappear" following the U.S. pullout.[6] The draft EIS public comment process also gave treaty opponents an opportunity to co-opt environmental language. Conservative congressmen seeking to preserve U.S. sovereignty over the Canal Zone had long used the sea-level channel's potential ecological threats for their own ends. Zonian interests now argued that the final EIS should focus more on "the human/community environment," by which they meant the livelihood and living conditions of the Zone's white U.S. residents.[7]

An esoteric debate over invasive sea snakes and starfish thus led to incongruous outcomes that spoke to the challenges of international, proactive environmental assessment. Marine biologists continued to promulgate the relevance of struggling naturalist disciplines to megaproject planning. U.S. environmental NGOs compelled the administration to pay closer attention to the environmental consequences of both the sea-level canal and of the transfer of the original waterway and Canal Zone to Panama.

During the intense campaign to win the requisite senate votes for the treaties, the State Department responded to environmentalist pressure for a stronger EIS by issuing an extraordinary statement designed to mollify NGOs that might oppose ratification. Only seven years after Lyndon Johnson's canal study commissioners dismissed environmentalism as a passing fad, NEPA had enabled environmental interest groups to exert significant influence in Washington. Yet they also risked contributing to the long legacy of U.S. imperialism in Panama by accepting the State Department's reassurances rather than using NEPA to hold the agency accountable for producing an incomplete EIS.

The Panama Canal Treaty EIS

President Carter's surprise announcement at the Yazoo City town hall on July 21, 1977, about the value of keeping U.S. options open for a future Panamanian sea-level canal, created a flurry of extra work not only for the treaty negotiators, but also for several executive agency employees. As the ambassadors raced against their August 10 deadline to strengthen the sea-level canal clause, officials debated how to meet environmentalists' demands for a full NEPA review. On August 1, Marion Edey, one of the still-unconfirmed members of the Council on Environmental Quality (CEQ), the White House office responsible for reviewing environmental impact statements, wrote a long letter to Stuart Eizenstat, the president's domestic affairs advisor, about the project's environmental background.[8] Edey was a well-connected member of the D.C. environmental lobbying circuit, having coestablished (with David Brower, the president of Friends of the Earth) the League of Conservation Voters in 1970. Although Department of Transportation officials had rejected the argument that the energy crisis justified reexamining the seaway proposal, Edey recommended proceeding with a "Quick E.O.P. [Executive Office of the President] State-of-Knowledge Review" to collate all the data from relevant agencies to determine whether to support Gravel's bill for a new study by the Corps of Engineers costing $7 million (the current equivalent of nearly $30 million).[9]

On the same day, August 1, Presidential Science Advisor Frank Press provided Carter with his office's summary of the sea-level canal ecology situation.[10] Having received the president's authorization to ask the National Academy of Sciences to conduct a new assessment, Press immediately conveyed the request to academy president Philip Handler.[11] It was a tall order—to assemble seven years' worth of work within eight weeks—but the academy agreed to sponsor a conference of experts in September and to prepare a report soon thereafter.

Press's office apparently acted without having consulted the CEQ, as Edey wrote him on August 23 to outline the many questions her office considered paramount for the academy study. Ironically, her letter mentioned that Katherine Fletcher, the domestic policy staffer who had nudged the CEQ back in April in response to environmentalist George Alderson's request, had suggested that the CEQ and the Office of Science and Technology Policy work together on the seaway issue.[12] The failure of two White House agencies to communicate in a timelier manner revealed the disorganized executive-branch approach to the seaway proposal, as well as the mounting pressure by the Washington-based environmentalist community to subject the canal treaty to NEPA oversight.

In the meantime, on August 5, representatives of the CEQ and several of the concerned environmental groups met in Washington with the assistant secretary of state for oceans and international environmental and scientific affairs, Patsy Mink, to discuss the Panama Canal Treaty's environmental impacts. State Department officials admitted they had not yet started the assessment.[13] But just a few weeks later, on August 29, the draft EIS announcement appeared in the *Federal Register*, inviting interested parties to submit comments within thirty days. The National Academy's new ad hoc group had not yet met, let alone issued its requested report to the president. Consequently, many of the eighteen federal agencies, nineteen private organizations, and seven individuals who responded to the Panama Canal Treaty draft EIS castigated the rushed job. Not only was the report short (forty-two pages, excluding the appendices), but the State Department violated the NEPA statute providing commentators forty-five days. The agency did reinstate the full period of public comment (that is, until October 13, 1977), but did not hold public hearings.[14]

Commentators criticized the draft EIS on several grounds, especially the superficial coverage of the sea-level canal. The Smithsonian Institution and the Center for Law and Social Policy (the public interest law firm that had litigated the Alaskan pipeline) took particular exception to the report's breezy acceptance of the Anderson Commission's conclusion that the sea-level canal posed acceptable ecological risks.[15]

The report also neglected the related issue of how supplementing the existing canal's reservoir might influence marine species migration.[16] Biologists influenced by the earlier seaway debate had passed resolutions calling on the Panama Canal Company to protect the waterway's freshwater barrier by not pumping in salt water to maintain adequate levels during periods of water scarcity.[17] After a record-setting drought in the winter of 1976–77 forced canal operators to restrict transits, however, scientists feared the proposal to augment the Gatun and Alajuela reservoirs with ocean water would gather momentum.[18]

Commentators called for the final EIS to provide detailed analyses about a variety of other issues. Panamanian decision makers and everyone concerned with the canal watershed needed much more information regarding the treaty's proposed joint oversight commission for protecting the canal watershed; the provision of financial aid to Panamanian institutions dedicated to conservation, sanitation, and the rational exploitation of natural resources; and past and present land and watershed management practices in the Canal Zone.[19] The notion of allowing Panama to develop its resources as Panamanians saw fit constituted an environmental parallel to the anticolonial rationale for the treaties, though it received little press attention.

Other than the Panama Audubon Society, the Gorgas Memorial Institute of Tropical and Preventive Medicine, and several Canal Zone civic councils, no Panama-based organizations submitted comments on the draft EIS. A U.S. nongovernmental organization, however, adopted an explicitly anti-imperialist approach to environmental management. International program director Michael Wright of the Nature Conservancy, a nonprofit known for buying private lands for preservation, declared, "The decision to preserve these unique Panamanian forests must ultimately rest with Panama." Having recently visited the country, Wright expressed confidence in the Panamanians' commitment to protecting the canal watershed from deforestation-induced erosion. Like other environmental groups, he recommended providing aid "to help mitigate the difficult economic choices such protection could involve." Wright praised the treaty overall for providing an "admittedly unintended" opportunity to promote environmental stewardship and cooperation throughout Latin America.[20]

The Panama Canal Company and Canal Zone government, the quasi-colonial partnership doomed to oblivion by the new treaty, did not see any such value, not surprisingly. The environmental quality committee representing both entities excoriated the draft EIS for not employing a broad enough view of human ecology. Because the Zone's U.S. citizens faced serious threats to their employment privileges and living conditions once Panama assumed jurisdiction, the final EIS, the committee said, should better attend to "the anticipated effects on this element of the human environment."[21] The presidents of five Canal Zone civic councils made similar points, and cast aspersions on the ability of Panamanians to act as responsible canal operators and watershed stewards.[22]

Likewise, treaty opponent Senator James Allen of Alabama used the draft EIS to bolster his case against ratification. Urging his fellow senators to "listen carefully to the recommendations of environmental groups," he delivered a speech on September 28, 1977, raising the specter of Panamanian-caused deforestation in the Canal Zone and the loss of environmental safeguards afforded by U.S. law. Because the draft EIS had also considered how Panamanian-imposed canal toll increases might affect environmental and economic well-being, Allen focused on the increased atmospheric pollution and higher consumer costs that would ensue if shippers of coal, petroleum, and other raw materials destined for U.S. markets avoided the canal in favor of overland routes.[23]

The final version of the EIS came out late in December 1977; it did not list any authors but identified William Mansfield III of the State Department's Office of Environmental Affairs as the contact person. The report retained a heavy, though not exclusive, focus on how environmental changes under the treaty might impact U.S. interests. It provided detailed data on the jobs and commercial and defense infrastructure at stake, and concluded that deforestation and insect-borne disease constituted "the potential environmental consequences which would appear to justify the greatest concern."[24] Should current environmental management practices discontinue, both phenomena would harm U.S. and Panamanian residents and economic interests. The loss of the Zone's forests would also threaten wildlife, including ten mammals and five birds listed under the U.S. Endangered Species Act of 1973.[25]

At the same time, the final text acknowledged Panama's dilemmas as a developing country and its resolve to address environmental problems. It cited, for example, the request by RENARE, the Agriculture Ministry's agency responsible for forests and national parks, for assistance in implementing a $20 million watershed management project.[26] A brief section on the archaeology and environmental history of the Canal Zone also made the point that its forests were far from primeval: "Though it comes as a surprise to most people, some evidence indicates that at the time of the Spaniards' arrival around 1500 A.D. the Indian population was so large that less of Panama was forested than was the case until very recently." Therefore, the conquistador Balboa probably "passed through planted fields and not through solid forest, in crossing the isthmus to 'discover' the Pacific."[27] Such use of scare quotes by non-Indigenous authors was unusual for the time, several years before the Columbus quincentennial increased public awareness of the precolonial Americas as community-managed landscapes. Otherwise, however, the final impact statement did not provide the in-depth historical evaluation of environmental management practices that several commentators sought.

The final EIS did at least provide more data on the sea-level canal. The report summarized the conclusions of the Anderson Commission, Battelle, CERIC, and the 1970 Biological Society of Washington symposium, and included the terse conclusion of the National Academy's recent update to the White House: while "the modest advances in knowledge" attained since 1970 could not be used to quantify the exact risks of ecological disruption to the Atlantic and Pacific biota, they justified the expense of implementing a barrier system designed to restrict species migration across the seas. The final EIS report also assured readers that the Panama Canal Company had no plans to pump seawater into its lakes, and that despite the treaty's invocation of PNEs, "there is no consideration of employing nuclear devices in the possible construction of a new canal."[28]

How the Academy Was Unmuted

The fall of 1977 was a whirlwind for advocates and opponents of the Torrijos-Carter Treaties. As the Carter administration embarked on its public relations campaign, the post-TAPS debate was heating up over building new west-to-east pipelines to distribute the surplus Alaskan oil, and the comments were rolling in on the draft EIS. The National Academy of Sciences was also hard at work; in a remarkable turnaround time after receiving the presidential science advisor's request on August 1, an ad hoc group chaired by Alfred Beeton of the Great Lakes and Marine Waters Centers formed. The Beeton Committee held a conference of two dozen experts on September 1–2, solicited comments from almost two dozen more, and issued its eleven-page report on September 28.[29]

Some of the biologists who had played major roles in the 1968–70 discussions served on the new committee, most notably Ira Rubinoff and Peter Glynn of the

Smithsonian Tropical Research Institute (STRI). Rubinoff, who became STRI's director in 1973, had since come out against the seaway proposal, calling instead for the construction of a third set of locks with saltwater pumps, tidal gates, and a "toxic barrier" to kill fouling and migratory marine organisms.[30] Other participants included Lawrence Abele, William Aron, John Briggs, C. E. Dawson, Sylvia Earle, Joel Hedgpeth, Meredith Jones, John McCosker, William Newman, C. Richard Robins, Richard Rosenblatt, Howard Sanders, Geerat Vermeij, and Gilbert Voss. Despite his rocky relationship with the scientific community, even John Sheffey participated, presenting on physical barriers to faunal mixing and on the Anderson Commission itself.[31]

The Beeton Committee report packed a lot of information into eleven pages, and provided the sea-level canal authors a fresh forum to explain why marine ecology mattered for maritime infrastructural development. Like the 1970 Mayr Committee on Ecological Research for the Interoceanic Canal (CERIC), the 1977 academy authors criticized the Anderson Commission for its notorious conclusion about the acceptable risks of adverse ecological consequences, and called for a barrier system to prevent the inevitable migration and colonization that would follow the opening of a sea-level waterway. But despite the passage of seven years, isthmian marine ecology had barely advanced. Few surveys had been conducted to identify the species capable of migrating, especially not in the deeper areas requiring special equipment and research vessels. Basic knowledge of marine parasites and disease organisms, even for commercially important organisms, remained sparse.

The Beeton report also called for a more complex approach to predicting the seaway's consequences for ocean life. The earlier controversy's focus on extinction events and charismatic species like the yellow-bellied sea snake and the crown-of-thorns starfish emphasized the direct effects of seaway construction at the expense of indirect effects on local marine communities. Just as the original Panama Canal had destroyed local mangrove forests, sea grass beds, and coral reefs when crews dumped masses of dredge spoil along the coasts, changes which in turn affected the members of multiple marine food chains, a sea-level waterway would affect nutrient dynamics, food webs, and species abundance by altering oceanic currents and sediment flows. State-of-the-art computer models, in concert with updated taxonomic information, offered promise for determining such localized effects. But otherwise, because the "imposing uncertainties" identified by the sea-level canal authors remained unresolved, the new academy analysis reiterated the importance of integrating marine ecology into any future engineering feasibility studies.[32]

Science Advisor Frank Press conveyed the report highlights to the president in early October, along with a mild warning: "I would recommend that you bear in mind the issue of potential ecological effects in your discussions and public statements and, as appropriate, acknowledge that the issue will require further detailed study."[33] Three days later, Carter faced the question of how committed he really was to the oil-crisis canal, when the director of the Office of Management and Budget, James McIntyre, requested his input regarding Gravel's bill to authorize the Army Corps of Engineers to conduct a new study. "As you consider it," counseled McIntyre, "the proposal should be viewed in the larger context of (a) the impact of the proposal upon obtaining Senate consent on the Treaties and (b) how the proposal would be received in Panama." On the plus side, he considered $7 million "a relatively small price to pay" for an updated assessment of the sea-level canal, especially if the energy crisis worsened enough to justify construction, now estimated at $6.2 billion. Yet a compelling counterargument could also be made: "Administration support for a sea-level canal study by the Corps—even though not a commitment to construct—will be strongly resisted by environmentalists who are concerned about potential adverse environmental and ecological effects from mixing waters from the Pacific and Atlantic Oceans, e.g., introduction of poisonous Pacific sea snakes into the Atlantic." After implicitly referencing the Beeton report, McIntyre emphasized the project's questionable economic returns, even in the context of increasing energy-transportation problems.

Carter's two options appeared at the end of the memorandum, under the heading "Presidential Decision." At some point, in his careful script, he checked and initialed the second choice: "Do not support legislation to authorize study."[34] It must have been a difficult choice for Carter, having raised so much fuss among the treaty negotiators just weeks earlier. Yet he must also have known that he had secured Gravel's vote for ratification, leaving him free to concentrate on other senators—a vital task given that public opposition to the treaties was running two to one.[35]

Most analyses of the ratification campaign focus on the rapid, unexpected rise of the New Right, and the immense pressure grassroots neoconservative groups exerted on undecided senators to vote against the treaties.[36] Environmentalist NGOs might not have possessed the resources to flood senatorial offices with millions of letters about the treaty EIS, but NEPA afforded them leverage by enabling them to sue federal agencies for producing inadequate environmental impact statements.[37] Although Carter's team had shown little interest in environmentalists' concerns about the treaty earlier in the spring and summer, that started to change in the wake of the September 6 telegram publicity. At one of the first post–signing ceremony ratification campaign events held by the White House, on September 15, 1977, environmental groups constituted six of the thirty-two civic organizations represented.[38] Carter's subsequent decision in early October not to authorize Gravel's bill for a new sea-level canal study probably reflected a deeper appreciation of the proposal's controversial history, as conveyed by the National Academy's Beeton report.

Coverage of the Beeton Committee findings by high-profile venues generated unflattering publicity for the administration and Article XII. *New York Times* science writer Walter Sullivan, who had covered the debate seven years earlier, published an article on October 10 that began, "The revival by President Carter of the proposal for a sea-level canal across Central America has evoked renewed concern among marine biologists about the effects of unimpeded access between the tropical Atlantic and Pacific."[39] The London-based *New Scientist* addressed the new scientific report as well as the antiseaway advocacy of Friends of the Earth. Describing the group's interpretation of the seaway as "a sinister move on the part of the oil lobby," the magazine quoted from a scathing editorial by President David Brower that appeared in the November 1977 issue of *Not Man Apart*, FOE's newsletter.[40] Brower's piece accused the Carter administration of "a breakdown in decision-making." Likely drawing on his close relationship with Edey, he excoriated Carter for promoting the project without having solicited CEQ's recommendations or directing an EIS to be conducted before the treaty talks began. Instead, days before the treaty signing, "an obviously inadequate draft EIS was released by the State Department," thereby preventing the administration from making an informed decision in accordance with NEPA.[41]

Brower alleged that Carter had responded "to the persuasion of a senator whose vote he wanted for ratification of the treaty," citing Gravel's public admission that he had met with the president to discuss Panama issues shortly before the Yazoo City pronouncement. In Brower's view, the Alaskan senator tried to circumvent Congress, the CEQ, and the public by persuading Carter to have the sea-level canal written into the treaty, thereby mandating a new government-funded feasibility study. The allegations gained a much larger audience when *New Scientist* republished his biting conclusion: "Proponents of the treaty have urged Friends of the Earth to be silent on the sea-level canal provision because it was put into the treaty only to satisfy Senator Gravel." Even so, argued Brower, "when one Senator can talk a president into a commitment that flies in the face of long-standing and well-documented environmental and scientific objections, we cannot stand by silent."[42]

More bad publicity addressed the ways in which megaprojects compounded the problems of tropical developing nations seeking to accelerate modernity. In December 1977, a widely distributed *Washington Post* article by a World Bank ecologist, Robert Goodland (the first person hired in that position), framed the sea-level canal as one component of a "triple threat to Panama ecology." The country also faced massive changes due to a proposed highway through the Darién Gap and the newly completed Bayano Hydroelectric Complex, which displaced thousands of Indigenous people in the eastern portion of Panama west of the Darién. The three huge projects embodied the challenges of reconciling economic development with respect for human rights and environmental quality.[43]

Much more was thus at stake than invasive sea snakes and starfish. Rubinoff and another Smithsonian official, Ross Simons, made that clear when the special assistant to the president, Joseph Aragon, paid them the compliment of asking for their take on the environmental issues raised by the Panama Canal Treaty. The two scientist-administrators urged the administration to provide technical and financial assistance for natural resource management, noting that "such an initiative would be welcomed from Panama as long as it did not appear to be an 'imperialistic' scheme." They also called for the joint commission specified in Article VI to include a robust scientific basis and to make the most of an opportunity for a new era of hemispheric partnership: "The tropics are being destroyed with extreme rapidity, a situation which in the long-term could have the same social and economic consequences as our current energy problems. We in the temperate zone should recognize that we cannot divorce ourselves from the tropics." As for the seaway, "perhaps the most emotional issue to environmentalists," Rubinoff and Simons advised Carter's team to assure them that no decision would be made in the absence of further scientific study.[44] Representatives of the most vocal environmental groups had also been calling for such assurance, though they would likely have objected to being described as emotional.

Reassuring the Antiseaway Community

As the administration's ratification initiative ramped up late in 1977, perceptions intensified that environmental organizations might mount a court challenge to demand a more thorough treaty EIS.[45] Although the State Department did not produce the comprehensive impact study environmentalists sought, officials agreed to develop a formal statement to accompany the release of the final EIS in December. The document, developed with input from environmental lobbyists, would clarify U.S. intentions regarding key points of diplomatic concern and thereby facilitate "support for ratification of the treaties by the environmental community," as a confidential memo by several of the major environmental players spelled out on December 14, 1977.[46]

A major figure who took credit for pushing this process along was David Ortman, who had received George Alderson's files when he took over as the research associate of the Washington office of Friends of the Earth. As he later asserted, "In the FOE D.C. files on the Panama sea-level canal are State Department memos expressing concern after calls I would make to the State Department, that the environmental community might scuttle the treaty efforts."[47] In the aforementioned November 1977 issue of *Not Man Apart*, Ortman wrote a long critique titled "Mingling the Two Oceans" that evinced considerable insider knowledge, which spoke to the close relationship between FOE personnel and CEQ member Edey. He also recognized the irony that "the biggest opponents" to the sea-level canal since the 1960s "were not the biologists or the environmentalists, but rather those people who are convinced that we should not give up the Canal Zone," a reference to Representative Dan Flood, Senator Strom Thurmond, and other adversaries of treaty reform.[48] On the other hand, Ortman did not seem to fear aiding such antienvironmental politicians; all that mattered was preventing the terrible seaway scheme from coming to fruition.

Ortman's analysis recapitulated the important point that "the possible ecological effects extend far beyond the sea snakes." Such threats included larger oil spills by the massive tankers that would use a sea-level channel, the need to deepen U.S. ports to accommodate these vessels, and the effects on the urban ports of Panama City and Colón of reducing traffic through the existing canal. Better solutions for reducing reliance on foreign oil existed than spending $10–20 billion on a huge public works project in Panama: "How much insulation and solar space and water heaters would the money spent on a sea-level canal buy? How many homes could be retrofitted to burn less oil?"[49] Ortman also challenged Gravel for ignoring the project's secondary environmental effects, overstating its military rationale, and underestimating its cost. His final theme echoed an argument Congressman Flood had made long before: "Besides Senator Gravel, President Carter, and Secretary of State Vance, who advocates for the colossal construction project? Well, engineering-industrial groups for one, manufacturers of heavy earthmoving machinery, dredging combines, and contractors."[50]

Ortman and fellow FOE staff were not the only ones who pressured the State Department. The Sierra Club, one of the oldest conservation groups in the United States, also played a big role. It had only begun expanding its international presence since 1970, yet had achieved great success using NEPA to challenge foreign megaprojects slated to be built with U.S. funds.[51] In particular, the club and three other organizations sued the U.S. Department of Transportation and Federal Highway Administration in 1975 to demand a better EIS for the proposed completion of the Pan-American Highway through the Panama-Colombia border. The plaintiffs sought a greater focus on preventing the spread of hoof-and-mouth disease (aftosa) from South American livestock across what had long functioned as a biogeographical barrier—the dense rainforests of the Darién. Because the disease threatened to kill off a quarter of the North American cattle herd—a $10 billion loss—it gained a lot of traction. A U.S. district court agreed that both the draft and final impact statements failed to adequately address aftosa, as well as the cultural survival of the affected Guna and Chocó tribes.[52] The rulings precluded construction of the 250-mile-long route for over two decades, and the Darién still remains the only incomplete stretch of the road from Alaska to Tierra del Fuego.[53]

Top representatives of the Sierra Club, FOE, and the Center for Law and Policy met with State Department officials three times in 1977 to address the treaties' environmental protection measures and the agency's compliance with NEPA (on August 5, November 18, and December 20).[54] "While we believe we have been able to allay their concerns," wrote Deputy for Panama Canal Treaty Affairs David Popper to Deputy Secretary of State Warren Christopher on December 21, "they would like a statement of assurances from you which they could use with their directors and members to win their support for the Treaties."[55] Actually, the environmentalists wanted Carter's secretary of state, Cyrus Vance, to deliver the message via a major speech.[56] Vance had a deep knowledge of U.S.-Panama affairs, having served as the Panama Canal's sole shareholder (as secretary of the army) and Lyndon Johnson's deputy secretary of defense during the 1964 Flag Riots. Instead of a high-profile speech by the nation's top foreign policy official, however, the environmental community had to settle for a glorified cover letter by the second-ranking foreign affairs leader, which was still not too shabby.

Christopher issued the "Statement on the Panama Canal Treaties and Environmental Protection" on January 12, 1978. The document declared that Article XII provided only "for a *study* of the feasibility of such a canal without making a decision or commitment that a sea-level canal will be built," which would include a full EIS addressing the reports of the National Academy. The text also downplayed the likelihood of nuclear excavation techniques ever being used, "both for environmental reasons and because of the terms of the nuclear test ban treaty."[57]

The language in Christopher's statement regarding Article VI, the section committing the two nations to respect environmental goals, was more questionable. He identified Article VI "as an integral part of the Panama Canal Treaty," a dubious claim given the pact's overriding focus on the operation, defense, and transfer of the waterway and associated lands and property. The statement also conveyed the State Department's intention that the Joint Commission on the Environment "shall have the staff and financial support it needs to be effective" and include among its U.S. members "leading science and environmental figures as well as others from the private and public sectors."[58] But because the treaty did not include such specific language, Congress would likely have to approve legislation to implement the plan.[59]

Regardless of what might happen in the future, Christopher's statement performed an important function for the antiseaway environmental community: it validated the 1960s-era sea-level canal authors and the political activists who took them seriously. Recalling the scientific demands for natural-history inventories of the isthmus, the document asserted that the U.S. and Panamanian governments would cooperate to collect baseline data "showing the current state of Canal Zone ecosystems, including air and water quality, marine life in the adjacent oceans, and flora and fauna." Another nod to the debate of the previous decade cited two older treaties as a legal precedent for U.S.-Panama environmental cooperation, including the one that established Barro Colorado Island, the original home of the Smithsonian Tropical Research Institute, as a nature monument in 1940.[60]

The Sierra Club was relatively pleased with the outcome, as articulated by its international committee chairperson, Nicholas Robinson.[61] The lawyer had a history of pressuring the State Department and other U.S. foreign affairs agencies to comply with NEPA; as he argued in 1974, Congress intended the law to extend beyond the territorial boundaries of the United States. That was not to say that NEPA permitted interference in the internal affairs of other nations. "Rather, the aim is to assure that the United States itself is never responsible for unanticipated environmental injury anywhere. NEPA provides a restraint on U.S. action, not on the actions of other countries." The law required all federal agencies, no matter their scope, to act "with as full an awareness as possible of their impact on the systems of the biosphere."[62]

Robinson criticized the State Department for waiting too long to initiate the Panama Canal Treaty EIS—it should have started during the Nixon administration, but at least senators could consult it prior to the ratification vote. While the agency had under previous leadership "flouted NEPA more than it followed the act's mandate," the Carter administration's actions represented "a solid step toward reversing that pattern." Robinson lamented the public lack of attention to the treaties' environmental dimensions, while making the ironic admission that the sea-level canal provision committed the United States "to so little" that the failure of environmentalists to achieve their desired language was "of little importance." As for the joint commission outlined by Article VI, environmental groups would have to ensure that the enabling legislation met the State Department's ambitious objectives. But first, of course, the pacts had to be ratified.[63]

By late January 1978, as the ratification campaign approached fever pitch, the sea-level canal option was becoming a white elephant. Despite strong support among some of their colleagues for it, the leading protreaty senators, Robert Byrd of West Virginia and Howard Baker of Tennessee, expressed to Christopher their willingness to delete it in order to "pick up votes," especially that of one senator, Robert Griffin of Michigan. Christopher cautioned them against this maneuver on the grounds that the article was in the national interest, and that any deletions would put the treaty at risk of further revision. Because Griffin appeared to be turning against the treaties anyway, Baker agreed to shelve the plan for the time being. When informed of these developments by Secretary of State Vance, Carter reiterated his support for the sea-level canal article: "It's important, I believe, to hold this in the treaty."[64]

As Carter wished, the sea-level canal clause remained, but it became an anticlimactic nonissue during the final months prior to the ratification votes in the spring of 1978. By early February, sixty-two senators had committed to support the treaties, twenty-eight opposed them, and ten remained undecided. To obtain the requisite sixty-seven votes, the administration made many last-minute concessions, including one that infuriated the Panamanians and almost derailed the whole process—a reservation introduced by Arizona senator Dennis DeConcini allowing the United States to use military force in Panama to keep the canal open, if necessary, after the year 2000. The Senate approved each of the pacts with one vote to spare; the Neutrality Treaty passed on March 16 and the Panama Canal Treaty on April 18 with identical roll calls.[65] It had been a bruising fight, requiring Carter officials to provide statements of reassurance not only to environmentalists fearing a future sea-level canal but also to reluctant treaty supporters fearing the loss of the perpetual unilateral right to defend the waterway.[66]

The Posttreaty Seaway Hearings

After the rancorous ratification campaign, Senator Mike Gravel remained convinced that a sea-level canal offered the best solution to modernizing the Panama Canal and to transporting North Slope oil from Valdez to tankers bound for the ravenous refineries of the East and Gulf Coasts. Although the Trans-Alaska Pipeline was only transmitting a few hundred barrels per day, energy analysts expected that number to skyrocket in the near future.

On May 4, 1978, the Senate approved moving forward with Gravel's bill to provide $8 million for a new commission to conduct a full NEPA review of the seaway project. Though he had previously ignored the risks of marine species exchange, the measure included a flora, fauna, and ecosystem inventory of the Panamanian isthmus and assessment of the ecological effects of marine species movement. The bill allocated three years for the commission to complete all its work, including the economic and technological feasibility studies.

The House of Representatives, where antitreaty sentiment remained strong, held three days of hearings to consider Gravel's plan in June 1978. Despite little enthusiasm for the proposal, one representative conceded, "I must admit that Alaskan oil and its importance to the entire Nation is a whole new factor, relatively speaking, in consideration of a sea-level canal."[67]

Gravel's testimony came back time and again to the present canal's technological obsolescence. In response to questions about the decreased annual transits, and to Senator Jesse Helms's claim that 98 percent of the world fleet could still go through the Panama Canal, Gravel used data he had commissioned to argue the opposite. Despite the drop in the number of ships, he argued that actual tonnage through the canal did meet earlier predictions due to the increase in average vessel size. However, because of its reliance on supertankers, the oil industry could no longer make the most of the Panama route, and thus the canal could only be said to have accommodated 42 percent of the world shipping fleet in 1977, a figure that would likely drop to 7.64 percent by 2000. Besides, Big Oil would not be the only beneficiary of a deep and wide new waterway; the U.S. construction, steelmaking, shipbuilding, dredging, coal mining, and West Coast timber industries would also profit. "Let us look at the economics of the future, and not the past. I would rather the Congress not fight about something that is obsolete," Gravel implored his colleagues, in spite of the arduous campaign to replace the archaic 1903 treaty.[68] He failed to see that for the antiratification forces, the original canal remained a living symbol of U.S. power, both in terms of past glories and future uncertainties in a post-Vietnam political culture.[69]

Gravel also tried to persuade his House colleagues that a new seaway would enable the navy's largest aircraft carriers to cross the isthmus at a moment's notice, rather than having to rush to a dangerous hot spot by way of Cape Horn: "You cannot put a price tag on it, because you do not know whether that 10 days would be the difference of [forestalling] nuclear holocaust in the world."[70] By invoking a national security nightmare in the vein of the Cuban Missile Crisis, he put a Cold War spin on the original military rationale for linking the oceans, the Spanish-Cuban-American War of 1898.

When grilled about the project's environmental effects, Gravel revealed his terrestrial bias. The opposition of California and Washington politicians and citizens to building new ports and pipelines to transfer Alaskan energy "to the bowels of this country" meant, Gravel said, that "we are going to be faced with what we are doing today, which is essentially carrying the oil around the central part of the Western Hemisphere, in buckets." Rather than fighting environmentalists over creating new inland pipeline systems—which would be prohibitively expensive regardless—why not make the most of the oceanic/isthmian route? Building the requisite infrastructure across the part of the continent that was only 50 miles wide (in Panama) as opposed to where it was 2,500 miles wide (across the transcontinental United States) offered the least damaging solution to conveying the vital cargo eastward. If the seaway caused marine species exchange or extinction, it was an unavoidable trade-off of economic prosperity: "It is an [inevitable] ecological displacement, over maybe some fish, or something else."[71]

Gravel's less garbled point that it would be difficult to overcome the environmentalist opposition to new transcontinental pipelines was significant given his own role in curtailing environmental critiques of the Trans-Alaska Pipeline five years earlier. His TAPS battle scars might have taught him a hard lesson about the persistence of environmentalist opposition, or he sincerely believed that investing in a sea-level canal made more sense than expanding the nation's oil pipeline infrastructure. Either way, it fit the description of Gravel's reputation in the Senate (which ended when he lost the 1980 primary election) as a "maverick" and "loose cannon."[72]

Subsequent witnesses discussed the many significant world changes that had occurred since President Johnson's Canal Study Commission had released its final report. But one thing that had not changed was the frustration of many biologists upon seeing the proposal reemerge with only token attention to baseline research. Two of the Smithsonian biologists who had lobbied for years for the natural-history inventories, administrator David Challinor and invertebrate zoologist Meredith Jones, testified on the second day of the hearings. When asked if the $3 million needed for a decade-long investigation would be worth it if the sea-level canal were ruled out on economic and national security grounds, they affirmed so by focusing on the value of the isthmus for addressing one of biology's most fundamental concepts, the origin of species. As Jones explained, "here we have a readymade workshop for determining what is a species, in that at one point there was a continuous fauna from the Atlantic to the Pacific." Analyzing how fish, urchins, and numerous other organisms had changed after being separated by the isthmian land bridge for some three million years would provide new insights for evolutionary biology.[73]

University of Miami fish biologist C. Richard Robins also testified on the importance of providing research funds, especially to continue surveying the Panamanian coasts and to analyze understudied existing collections of tropical oceanographic animal specimens. Robins had contributed to the infamous Battelle report that downplayed the threat of faunal mixing, and still seemed to harbor resentment against the biologists who had dismissed his team's labors. In addressing the importance of providing opportunities for non-U.S. researchers, he made a not-so-subtle dig at the Smithsonian Tropical Research Institute: "On occasions when my colleagues and I have visited the University of Panama or have talked elsewhere to their scientists, we have always heard that they felt left out of scientific activities in their country by the Americans."[74] STRI, for which the Panama Canal Treaty included a special protective provision, had indeed attracted criticism for not including enough opportunities for local students and scientists. Under Rubinoff's leadership, STRI increasingly expanded its Panamanian employment, training, and education programs, actions which helped the institution negotiate additional agreements with the government of Panama in 1985, 1997, and 2000 to continue operating in the former Canal Zone.[75]

Although no other researchers presented at the hearings, some harsh assessments by scientists made their way into the official record. William Newman's acerbic 1972 reflection on the relationship between the Anderson Commission and CERIC appeared as "Attachment D," and a private letter to Beeton from John McCosker, one of the few biologists who had published on the Panama Canal as an avenue of faunal exchange since the 1960s controversy, conveyed his distress regarding the megaproject's revival: "I am disappointed, chagrined actually, to discover the resurrection of the Sea-Level Canal concept. I would have thought the previous exercise was satisfactory to demonstrate the futility of such an experiment before adequate baseline data had been collected."[76] This sense of disgust was captured in a separate publication by Gilbert Voss, one of the University of Miami biologists who had angered the CERIC scientists, in which he exclaimed, "not another sea-level canal!"[77]

Representatives of the environmentalist and conservation communities testified on the third and final day of testimony. Ortman of FOE delivered a sharp denunciation of the sea-level canal on several fronts, and his *Not Man Apart* article was entered into the record along with the telegram delivered to the White House on the eve of the Torrijos-Carter signing ceremony. A representative of the Izaak Walton League of America, a venerable organization of about fifty thousand anglers—almost twice as many members as FOE—also spoke. Both argued that Gravel's bill did not afford adequate attention to ecological research due to its three-year timeframe, far less than biologists had requested.

One of the most interesting testimonies came from John Sheffey, the former executive director of the Anderson Commission. He remained convinced that a sea-level canal would be a better investment than the third-locks expansion project, which would be rendered obsolete as ship sizes once again increased. Yet he did not think it necessary to spend $8 million for a comprehensive new study. Instead, Sheffey supported an alternative bill by Representative John Murphy of New York to spend two years and $1 million to determine the feasibility of a new interoceanic waterway in accordance with Article XII; *if* the president then decided to proceed with the project, Congress would provide an additional $2 million to conduct an EIS in accordance with NEPA.[78]

When asked why the environmental study should not begin immediately, Sheffey showed that he still conflated the scientific sea-level canal authors with political activists, and that he expected scientific studies to convey a high degree of certainty: "My judgment, based upon the knowledge I acquired from the environmentalists during the 1965–1970 studies, simply is that they cannot tell you, over any length of time, definitively, whether or not there will be environmental harm to the ocean populations by making the canal." He continued, "There is no possible way to make a laboratory model similar to the oceans, nor is any reasonable length of time sufficient to reliably predict what will happen." The Rubinoffs and others had conducted breeding experiments in tanks and had towed species through the canal to start to provide a quantitative basis for predicting the effects of marine species mixing. For Sheffey, however, the fact that "some of the impacts will not be known for 50 or 100 years" justified long-term ecological risks in favor of immediate economic benefits. He considered the Erie Canal a case in point: even if people had known in the nineteenth century about the damage invasive sea lampreys would cause to the Great Lakes ecosystem, it would still have been worth building.[79]

Sheffey saw no reason to waste money on research for a megaproject that might not ever come to fruition—in other words, he did not understand the point of NEPA. As he saw it, once the president had decided for economic, military, and political reasons to build a sea-level canal, scientists would have plenty of time to conduct baseline ecological studies during the fifteen-year process of negotiating, planning, and building it. Even the needed geophysical studies, including in-depth analyses of subsurface geology, slope stability, and the hydrodynamics of ships moving through confined waters in variable currents, could await executive consent. When asked by an incredulous congressional staffer if he would support a decision to build a seaway without having completed a geophysical analysis, Sheffey replied, "Sure. We know we can solve the problems. There are not any unsolvable problems that could develop. Slope stability is merely a matter of excavation."[80] That he was as untroubled by the risks of triggering landslides as by unleashing oceanic bioinvasions exemplified the high-modernist mentality that an advanced state could manage any unintended consequence, ecological or otherwise.

At the hearing, Sheffey also provided deeper insight into President Johnson's reasons for supporting the sea-level canal, or at least his perceptions thereof. Once built, its relative invulnerability to sabotage would preclude on-site U.S. defense forces, and its comparative ease of operation and maintenance would enable the host country to quickly assume day-to-day operations. In the meantime, during the many years of construction, the waterway would sustain U.S. hegemony in the region: "We thought that Panama, in return for the huge investment in Panama, would give us the right to build and control a sea-level canal for a longer period than we could continue to control the lock canal. The economic case for it, as you see in our report, was marginal; the political and military cases for it were quite good."[81]

Sheffey had made a similar, controversial point regarding the Anderson Commission in 1970, when he let slip classified information about the group's recommendation to the president. As Representative Flood had fulminated to his colleagues, Sheffey told a *Wall Street Journal* reporter that a major purpose of the sea-level canal was to achieve "excellent treaty relationships" between the two countries and to end the clashes over canal operation and sovereignty.[82] At the 1978 hearing, a congressional staffer followed up by asking if Johnson's proposal was "simply a gimmick to enable us to maintain a U.S. presence in the Canal Zone because our policy makers had lost the determination to hold onto the existing canal." Sheffey bristled at the word *gimmick*, replying that Johnson had promoted the project "for the same reasons that the Senate has now ratified the new Panama treaties"—to facilitate the eventual transfer of the original canal to Panama. Johnson's vision of the future seaway had held indubitable benefits for the United States, but that did not make it a scam; as Sheffey explained, the president had been advised that a seaway was "probably feasible and could facilitate longer U.S. tenure of a canal and reduce the military risk of our ultimate departure from the Isthmus."[83]

In a separate interview conducted in 1979 by former ambassador to Panama William Jorden, Sheffey elaborated on the technological context of the seaway proposal: "Johnson really made this decision, I believe—the decision to negotiate [new treaties with Panama]—I believe under the very strong conviction that a very inexpensive sea level canal by nuclear explosives was in the cards." Fifteen years after the Flag Riots of 1964, Sheffey still smarted from having believed the hype about PNEs. As he confided to Jorden, the Plowshare physicists Edward Teller and Gerald Johnson had oversold the nuclear canal to Johnson's administration and many members of Congress, persuading Sheffey to change his life and career so as not to miss out on such an exhilarating endeavor.[84]

As for his argument for authorizing only a limited new study of the sea-level canal's feasibility, Sheffey emphasized the long history of commissions that had investigated the canal question. "The isthmus has been studied for 300 years, and our study was an updating, in effect of the [19]47 studies" by the Panama Canal Company, which in turn had built on studies dating back to 1906 and earlier. "They all were built on the past, and there is not a lot of new knowledge to be acquired. The new things are economic and political, not technical."[85] It was a remarkable revelation of how quickly the promise of nuclear excavation technology had faded.

The economic and political circumstances of the 1970s did indeed differ significantly from those of the earlier periods during which powerful stakeholders had invoked the sea-level canal as an ideal solution to the problems of isthmian transportation, and U.S. control thereof. Yet Sheffey employed a very narrow conception of what amounted to "new knowledge." Each phase of the controversial proposal generated novel understandings of nature and of how to engineer nature for diverse human goals. These insights in turn transformed anew the ideas of progress underlying the vision of the seaway of the future, giving rise to multilayered repercussions that would influence megaproject decision-making and statist environmental management in unforeseen ways for years to come.[86]

Conclusion

Sheffey misidentified the scientists who contributed to the 1965–70 sea-level canal discussions as environmentalists, yet by the time of the 1978 hearing, two of the most prominent biologists had embraced an activist role. Rubinoff spoke out against the project in an article that appeared in a 1975 volume associated with the United Nations Conference on the Law of the Sea, a nine-year-long forum that led to an international agreement governing multiple aspects of marine resource use.[87] More broadly, as STRI's director from 1973 to 2008, Rubinoff championed many initiatives to elucidate and protect Panama's marine and terrestrial biodiversity and to integrate the institute into Panama's scientific and educational community.[88]

Edward O. Wilson, the Harvard biologist who served on the National Academy's first canal committee, also inched toward advocacy in the succeeding years. He cited the sea-level canal "as an example of the worst thing that biologists might let slip by them" in a 1974 *Harvard Magazine* article that he later described as his "first venture into conservation activism." He also called it his first explicit application of island biogeographical insights to conservation planning.[89] To build a waterway capable of mingling the Atlantic and Pacific biotas "would be playing ecological roulette with all cylinders loaded."[90] Since then, the theory of island biogeography, which he codeveloped prior to the canal debate, has become what Wilson proudly calls "a foundation of modern conservation biology," a mission-oriented discipline devoted to maintaining and restoring biodiversity.[91]

The renowned evolutionary biologist Ernst Mayr, who worked with both Rubinoff and Wilson at Harvard and who chaired CERIC, did not participate in the 1970s-era controversy, nor did he address it in his later books on the history of biology. When Mayr died in 2005 at the age of one hundred, Rubinoff paid homage to his forgotten leadership of the National Academy committee. A journalist exaggerated Rubinoff's words, writing, "If it weren't for Mayr's tenacity, the proposed canal would have destroyed 3 million years of isolated evolution."[92]

The story was, of course, not so simple. Many other stakeholders during the decade after Mayr stepped away from the debate worked to ensure the government paid as much attention to the megaproject's ecological effects as to its economic, military, and geopolitical ones. The National Environmental Policy Act of 1969 enabled citizen environmentalists to mobilize the insights of biologists in powerful, unexpected ways. Adapting to the NEPA requirements entailed a steep learning curve for the State Department and White House. The hurried, uncoordinated response to the environmental groups protesting the sea-level canal provision of the Panama Canal Treaty cast in sharp relief the rapid evolution of NEPA-influenced environmental management during the "environmental decade."

The Carter administration spent a mountain of political capital to actualize the Panama Canal Treaties for both idealistic and pragmatic reasons. Transferring ownership of the canal and buffer zone righted the wrongs inscribed in the unjust 1903 treaty and eradicated an offensive relic of colonialism by recognizing Panama's sovereignty over all its lands and resources. At the same time, the treaties ensured continued U.S. *access* to the interoceanic waterway and the *right* to send in troops to defend it if threatened.[93] Despite all the blowback suffered by protreaty forces for their supposed surrender of an invaluable asset, the United States retained a significant stake in its long-term investment.

Lesser known provisions of the Panama Canal Treaty committed the United States and Panama to study and negotiate a sea-level waterway and to coexecute the agreement in environmentally sensitive ways. Carter and his diplomatic team sought to promote a more equitable relationship between the superpower and the small yet strategic nation, though they stopped short of immediately transferring all U.S. rights over the canal and its watershed to the Panamanian government. They also precluded Panama from negotiating with other wealthy nations to build a better seaway during the treaty's twenty-two-year life span, and did not ensure that the treaty EIS analyzed Panama's options for managing the canal watershed. At the same time, administration officials scrambled to prevent U.S. environmentalists seeking a stronger EIS from using NEPA to file a lawsuit against the State Department, an action that would have attracted adverse publicity and provided even more ammunition to conservative antitreaty interests.

Against all odds, the isthmian sea-level canal idea remained viable after both the demise of PNEs and the ratification of the new agreement with Panama. The seaway proposal survived because it performed important kinds of work for many different stakeholders in the context of the mid-to-late 1970s U.S. energy crisis. It provided an option for modernizing maritime transportation, for incorporating environmental values into public works planning, for prioritizing neglected forms of biological research, for considering the proper scope of extraterritorial environmental protection, and even for upholding remnants of the imperialistic status quo. Far from being a mere gimmick, the sea-level canal proposal provided powerful visions of alternative futures for those who skillfully deployed it.

Conclusion

Remembering the Unbuilt Canal

Environmentalists did not kill the sea-level canal, neither in 1970 nor in 1978. In each case, the U.S. government put the brakes on the megaproject because the fiscal, national security, foreign affairs, and domestic benefits did not converge at the right political moment to justify further investments of capital and energy. Chairman Robert Anderson of the Canal Study Commission had outlined the stakes back in March 1966. At a meeting attended by the Johnson cabinet's top-ranking members—the secretaries of state, treasury, commerce, and the army; the deputy secretary of defense; the chairmen of the Atomic Energy Commission and the Joint Chiefs of Staff; and the president's assistant for science and technology—Anderson explained that while the economic, defense, and political arguments for a new sea-level waterway were each insufficient to justify the endeavor, "the three together could prove sufficient to warrant its construction."[1]

During most of the 1960s, the experimental technology of peaceful nuclear explosives lent credence to the idea that the exorbitant expense of cutting a full channel across the Continental Divide could be reduced to a reasonable level—especially in concert with the diplomatic and military advantages of containing Panamanian demands to end U.S. control of the lock canal and surrounding enclave. An eventual streamlined seaway held the promise of satisfying Panamanian aspirations for full sovereignty while maintaining U.S. access to the Zone's strategic locale and assets for the foreseeable future. By the time the Anderson Commission submitted its final report in 1970, however, presidential concerns about violating the Limited Nuclear Test Ban Treaty had prevented Plowshare physicists from conducting all the needed PNE experiments. A large portion of the preferred route would not hold up to nuclear dynamite anyway, according to the geological surveys. Environmental scientists seeking a share of the feasibility study funds did succeed in drawing attention to the potential negative consequences of seaway-induced marine bioinvasions, and to the commission's questionable conclusion that such phenomena carried acceptable ecological risks. However, the demise of cheap nuclear construction methods, combined with the loss of momentum for U.S.-Panama treaty reform and the nation's huge budget deficits caused by the Vietnam War, played the major role in undermining high-level support for the 1960s iteration of the sea-level canal.

Later in the 1970s, in the context of increasing energy costs for consumers, of higher economic benchmarks for constructing critical energy infrastructure, and of the new treaty agreements with Panama, President Carter perceived that the sea-level canal's moment might have come. He secured a treaty option for the two nations to commit to a new study of a future seaway meant to transport Alaskan oil eastward. But he did not anticipate the intensity of opposition by environmentalist groups, who now possessed the ability to challenge public works planning via the National Environmental Policy Act. Using the ecological and evolutionary insights of the 1960s biologists who had called for extensive preproject baseline research, several Washington-based environmental NGOs urged the president to abandon even the study of such a waterway (a goal that not all the biologists shared). Carter's team scrambled to show respect for his environmentalist constituency by ensuring that the State Department complete an environmental impact statement for the Panama Canal Treaty. When activist groups criticized the final EIS as insufficient (for reasons that related not only to the sea-level canal per se), the administration tried to conciliate them by issuing a statement of reassurance. It worked, thereby avoiding an embarrassing NEPA challenge to the treaty during the grueling ratification campaign of 1977–78.

Yet that was still not the end of the sea-level canal story. Following the ratification events, the seaway proposal simmered on the back burner of U.S. economic and national security policy-making, especially as Japanese interests sought to assert their rising influence in the Western Hemisphere.

The Tripartite Hydrocarbon Highway

Even after the Article XII fiasco, and after Congress rejected the bills proposed by Senator Gravel and Representative Murphy to conduct a new set of feasibility studies, President Carter remained intrigued by the promise of a seaway for conveying Alaskan oil to the energy-starved East Coast. As the hearings to enact enabling legislation for the Panama Canal Treaties heated up after the 1978 midterm elections, environmental groups renewed their efforts to convince Carter to change course. President David Brower of Friends of the Earth, along with the directors of ten other influential NGOs, sent the president a letter on January 30, 1979, asking him to follow through on the State Department's intentions for the joint environmental oversight commission of the existing canal and to oppose any further congressional efforts to authorize a new study: "We see no need for haste and recommend that a decision concerning a Sea-level Canal study not be made part of the current implementing legislation."[2]

Most members of Congress appeared to be on their side; when the House Committee on Merchant Marine and Fisheries asked the congressional Office of Technology Assessment to analyze the Panama Canal Treaty EIS, the parties agreed in the summer of 1978 not to address the sea-level canal. The Office of Technology Assessment staff did, however, consult with four of the biologists who had contributed to the 1960s debate (Rubinoff, Jones, Robins, and Sanders) and noted their concerns about the EIS's lack of attention to the consequences of pumping seawater into the existing canal during droughts.[3]

The Panama Canal Act of 1979 did not authorize funds for new seaway feasibility studies, but it did echo the language of Article XII committing the two nations to study the issue. Otherwise, the congressional debates over implementing the treaties focused on military and economic issues. The enabling laws took effect just three days before the treaties entered into force on October 1, 1979, the first day the Panamanian flag flew alone over the former lands of the Canal Zone.

The possibility of cutting a new canal west of the Zone revived yet again in the new decade. Although the 1970 report of the Anderson Commission had generated revilement in the United States, the Japanese read it eagerly.[4] By 1980, Japan's postwar economic transformation was nearly complete, and it had embarked on a series of overseas development projects to secure strategic resources and its status as a global economic behemoth. The country's thriving economy would benefit enormously from a second Central American waterway, as the president of the Japan Chamber of Commerce and Industry and former head of Nippon Steel, Shigeo Nagano, explained: "We can bring through the canal at far cheaper cost grains from the United States Midwest, coal from West Virginia, oil from Venezuela, iron ore from Brazil."[5]

The estimated costs of building a sea-level waterway along the Route 10 site west of the Canal Zone had ballooned to $20 billion (including $8.3 billion for construction alone). But having a channel measuring 650 to 1,300 feet wide and 110 feet deep appeared worth it to the Japanese, whose ships accounted for over one-third of the Panama Canal's traffic. The lock canal could not accommodate vessels larger than 40,000 tons, and thus a seaway large enough for 300,000-ton tankers—designed to carry raw materials going to the island nation and automobiles and consumer electronic goods leaving it—would represent the pinnacle of Japanese technological mastery.

Japan's shipbuilding industry had exploited the 1967–75 closure of the Suez Canal by developing very large and ultra-large crude carriers and, later, specialized dredging technology for widening the Suez.[6] During the 1970s, the Panama Canal had gone from being able to accept 90 percent of the world's ships to less than 40 percent, and the number of vessels passing through had declined from 15,500 in 1970 to 13,200 in 1976 to 12,000 in 1978. Panama likewise had a great interest in not allowing the canal to become obsolete once it acquired full control in 2000.[7]

As the Panama Canal's wealthy, second-largest user, Japan commanded the attention of the U.S. and Panamanian governments. In January 1980, several Japanese bankers and businessmen visited Panamanian president Aristides Royo to discuss a $30 million feasibility study.[8] Five months later, in a Washington meeting with Prime Minister Masayoshi Ohira, Carter joked about the United States providing the engineering and equipment and Japan providing the money for a new sea-level waterway. He concluded that while it did present some environmental problems, they did not seem insurmountable.[9]

As in 1977, Senator Mike Gravel served as "the main catalyst behind this effort," as an administration official informed the president's assistant for national security affairs, Zbigniew Brzezinski, in March 1980. Anticipating that once again Gravel "might do an end-run to the President on this issue," Carter's lieutenants sought to manage the situation by "giving him a little bit more information" about the plan to inform the Panamanians and Japanese of the president's interest in a tripartite feasibility study.[10] The Alaskan senator had traveled around the world seeking financing opportunities for his passion project, even visiting Japan with Panamanian diplomat Eduardo Morgan, who declared, "If the feasibility study is positive, then the sea-level canal project cannot be stopped." Gravel shared his thoughts in a 1979 interview with William Jorden, who as ambassador to Panama had played a major role in brokering the canal treaties and who was writing a book on the subject. "Had there been strong support either from the Panamanians agitating or the White House, I would have got legislation last year [1978] that would have brought about the study," insisted Gravel. It was still not too late: "I feel that if we made a 'go' decision in '81, that it could be completed by '87."[11]

Yet no one else in Congress shared his enthusiasm, leading Gravel to a sad prediction: "And so it will just languish around until about 1990. Everybody will wake up and recognize it is highly obsolete and say, well, hell, we ought to do something, and we will have missed 15, 20 years."[12] Jorden later interviewed Deputy Secretary of State Warren Christopher about whether he foresaw a sea-level canal being built in the next twenty years. Christopher replied that it depended on how much Japan would want to contribute and pointed out that Gravel's defeat in the 1980 Alaska primary had removed from Congress the one member most committed to the issue.[13]

Following the May 1980 Carter-Ohira meeting, Japan, Panama, and the United States agreed to begin discussions on how to implement the tripartite study. The deliberations did not begin, however, until the administration of Ronald Reagan, who defeated Carter in the November 1980 election. By the time representatives of the three nations met in 1982, another major change had occurred: the Panamanian government had contracted with a U.S. company to build a pipeline near the Costa Rica border to transport North Slope crude oil eastward. Completed in 1981 with private financing, the eighty-one-mile-long Trans-Panama Pipeline proceeded without any state-mandated environmental review.[14] (By contrast, after five years of regulatory delays and environmentalist lawsuits, the U.S.-based Sohio oil company canceled its proposed $1 billion California-to-Texas pipeline and port in 1979.)[15]

The Trans-Panama Pipeline dimmed but did not kill the sea-level canal; following six meetings, the three nations established the Commission for the Study of Alternatives to the Panama Canal in 1985. Like the Anderson Commission of 1965, it had an ambitious five-year agenda to analyze the impacts of a sea-level canal, modifications to the present one, and overland options such as pipelines or container-rail transit. Unlike the previous group, however, the new commission was explicitly tasked with determining the environmental and social, as well as the economic and political, impacts of such alternatives.[16]

Yet the questions raised in the 1960s about nonnative marine species exchange remained inchoate, underfunded, and unfamiliar, even to its Japanese boosters.[17] When the U.S. ambassador to Japan, Mike Mansfield, asked Shigeo Nagano about the sea-level canal's environmental problems, Nagano replied: "We have no intention of using nuclear explosions in the construction." Mansfield, a former Democratic senator from Montana (who had retired just before the canal ratification votes), explained, "I'm asking about what will happen when the waters of the Pacific and the Atlantic oceans flow together and the marine organisms which have been separated for hundreds of millions of years mix together." Nagano was dumbfounded, a sign that the scientific and environmentalist communities still had much work to do in Japan—and elsewhere.[18]

Ecological sensitivity was also in short supply at a 1986 international gathering in Anchorage, Alaska, devoted to megaproject planning. Organized by a Swedish think tank, the International Federation of Institutes for Advanced Study, the conference featured presentations on such projects as a moon city, a dam across the Bering Strait, an Alaska-Norway ice highway, a river diversion network from the Pacific Northwest to thirty-three U.S. states and Mexico—and the Panatomic Canal. Then in his late sixties and working as a private consultant, John Sheffey reminded the audience of the benefits of nuclear dynamite, while conceding that a thirteen-megaton charge "would rattle windows 105 miles away." Regardless, political obstacles now posed "insurmountable problems" to nuclear geoengineering. Likewise, laughed the ice road proponent, the biggest hurdle would be having to file an environmental impact statement.[19]

After Reagan officials of the Council on Environmental Quality and the Environmental Protection Agency declined to help draft the terms of reference for the tripartite seaway study commission, the State Department asked the Smithsonian Institution to step in. The Smithsonian representatives remained focused on the need for a biological baseline survey because the isthmian coasts and oceans remained woefully underresearched: "Everything that follows must be founded upon a firm taxonomic and ecological base. Only when this is done, can we properly assess the possible environmental effects on an interoceanic canal or any other alternative."[20]

The needed scientific work remained unfinished, however, and by 1990 the prospects for the sea-level canal faded once again.[21] The Japanese stock market crash of 1989 led to a "lost decade" of growth and the end of its Lessepsian ambitions for a second Panama Canal.[22] Moreover, the U.S. invasion of Panama in 1989 to remove the dictator Manuel Noriega—whom officials accused of threatening the Panama Canal's neutrality—refocused attention on whether the nation could manage the original waterway within a decade, let alone build a new one. After the 1979 transfer of the Zone lands to Panama, many of the embittered Zonians who remained in the country blasted the Panamanian government for not cutting the grass properly, and reminisced about the Zone's similarity to "a beautifully manicured golf course," a function of the Panama Canal Company's well-funded mosquito-control practices and large groundskeeper force.[23] More substantively, Panama had to deal with the problem of cleaning up unexploded ordnance left behind by the U.S. military and with implementing new forms of canal watershed management.[24]

Challenges of Postcolonial Environmental Management

Despite the challenges of the twenty-two-year-long transition, the 1999 transfer of the canal proceeded smoothly, earning the Panamanian government praise for doing a far more effective job of operating it than the United States.[25] When it came to the long-postponed issue of modernizing the waterway, however, environmentalists criticized the Autoridad del Canal de Panamá (ACP), the government agency that replaced the Panama Canal Company, for replicating the undemocratic, heavy-handed decision-making processes that had prevailed previously. In 2006, the government held a referendum on whether to add a third lane of locks (measuring 180 feet wide and 60 feet deep, as opposed to 110 feet wide and 42 feet deep) at a cost of $5.25 billion, a plan related to the one advocated for decades by Representative Dan Flood. The decision grew out of several 1990s-era meetings that deemed a sea-level canal too expensive.[26] Although the "Sí" (Yes) measure to expand the existing channel passed by an overwhelming 80 percent, only 40 percent of eligible voters participated, and opponents expressed concern that corruption and costs would spiral out of control.[27]

Allegations that officials manipulated the voting process to ensure a favorable outcome, and the fact that the ACP did not complete the required EIS until nine months after the referendum, led to strong criticism from local and international observers. In the words of environmental legal scholar Carmen Gonzalez, "the EIA [environmental impact assessment] process was reduced to an empty ritual, a technical justification for a decision made at the highest levels of government and subsequently 'approved' in a 'democratic' referendum rather than a tool to inform and enhance public and governmental decision-making over Panama's single most important resource." Although Panama had sought for decades to escape the oversight of the United States, the colonial construct created by the canal remained palpable. As Gonzalez argued in a 2008 assessment, the ACP promoted a technocratic rather than democratic model of environmental impact assessment—a process that privileged compliance with preestablished regulatory standards rather than public involvement.[28]

The ACP completed the massive project over budget and two years behind schedule, in 2016. To address the problem of the canal's dwindling water supply, the engineers devised an innovative system of water-exchanging basins that enables the locks to reuse up to 60 percent of the water, rather than washing it all into the sea. However, pressure on the Gatun and Alajuela Lake water storage system remains high, especially in drought years. The expanded waterway's environmental effects are by no means confined to Panama, since ports around the world have deepened their harbors and made major infrastructural changes to accommodate the NeoPanamax vessels, whose carrying capacity is three times that of the previous generation of Panamax ships.[29]

Environmentalist concern and scientific interest in the Panama Canal as a model system for testing predictions about tropical marine invasions revived during and following the third-locks expansion of 2007–16.[30] Resulting studies, many funded by the Smithsonian, have confirmed that while most marine species cannot tolerate Gatun Lake's fresh water, hull fouling or ballast water may have facilitated recent invasions of macroinvertebrates and as yet undetected species.[31] Related experiments have elucidated the apparently greater ecological resistance of the tropics to marine invasive species than the temperate zones.[32] Yet researchers are still struggling to understand the fundamental question posed by the sea-level canal authors—how nonindigenous marine species manage to expand their range into new environments once transported there by human-mediated processes.[33] These questions are more urgent than ever; with the Panama Canal's doubled capacity for shipping traffic, marine species introductions will likely increase worldwide, especially in receiving ports of the U.S. Gulf and East Coasts.[34]

Atlantic and Pacific marine biotic mixing has also begun to escalate via the Bering Strait as the Arctic sea ice melts. Russian and other stakeholders began investing billions of dollars to develop a "Suez of the north," but little if anything for environmental impact assessment.[35] Smithsonian biologists have led efforts to address this research-policy gap, using language reminiscent of the sea-level canal controversy: "Reconnection of the Pacific and Atlantic Ocean basins will present both challenges to marine ecosystem conservation and an unprecedented opportunity to examine the ecological and evolutionary consequences of interoceanic faunal exchange in real time."[36] In this current moment of environmental crisis and intellectual opportunity, the conversation opened by the sea-level canal biologists fifty years ago remains deeply relevant.

The first official vessel to transit the new locks, an enormous Chinese vessel packed with almost 9,500 containers en route to the Pacific, heralded a new geopolitical era. Other milestones followed with the first U.S. liquefied natural gas (LNG) shipment to China in August 2016 and the first instance of three LNG tankers transiting the canal on the same day, in April 2018. Transported in NeoPanamax vessels, U.S. LNG had not been profitable in Asia prior to 2016. Since the canal expansion, the industry has expanded rapidly, developing new facilities and terminals to facilitate the booming commodity, a result of the concomitant hydraulic fracturing shale gas revolution in the United States.[37]

The LNG gas boom played out against the backdrop of plans to build the largest canal in history, a new Atlantic-Pacific link spanning Nicaragua. In 2013, the country's government announced that it had granted a fifty-year concession to a Chinese billionaire to finance the $40 billion, 170-mile-long channel, leading to speculation that it was a joke.[38] But after President Daniel Ortega claimed in December 2014 that construction had already begun, the international scientific community mobilized against the megaproject. Environmentalists and biologists from the Nicaragua Academy of Sciences, STRI, and elsewhere around the world raised grave concerns about the project's effects on water resources and biodiversity, and urged the government to consider the environmental guidelines and human rights laws that now govern infrastructure decision-making in many countries.[39] The protests led the Nicaraguan leadership to authorize an environmental and social impact assessment, though critics deemed it superficial. The country's highest court paved the way for the work to resume by dismissing the last environmentalist challenges in 2017. By then, however, the Chinese concessionaire had lost most of his telecom fortune and Panama had reestablished diplomatic and economic relations with China (and dropped them with Taiwan), intensifying assumptions that the plans had crumbled.[40]

Since Nicaragua announced its plans for the Grand Canal, other nations have upped the ante by proposing even larger maritime highways. Proposals to cut supersize canals through Thailand and Iran have raised concerns that the economic, national security, and political rationales will drown out environmental discussions.[41] The Nicaragua outcome thus holds high stakes for mediating hubristic plans for carving through continents to suit powerful interests. More broadly, conservation biologists and environmental and human rights activists argue that moving beyond the "global era of massive infrastructure projects" that deliver enormous benefits only for the lucky few requires convincing planners and investors to apply realistic assessments of ecological, social, economic, and political risks, and to otherwise resist perpetuating "megaproject imperialism."[42]

Focusing on the immediate economic and geopolitical payoffs of proposed megaprojects, however, is a hard habit to break. When he overrode his own experts' recommendation to build a sea-level canal across Panama in 1906, President Theodore Roosevelt cited the need to canalize the isthmus as soon as possible, and left the question of keeping up with expected increases in ship sizes to future generations. The Panama Canal's operators did begin building a third lane of locks in 1939, but after World War II scuttled the project, U.S. officials spent the rest of the century debating whether it made more economic, military, and political sense to engineer a new channel at sea level. When Panama obtained control in the twenty-first century, it chose to resume the 1939 endeavor at a cost of $5.25 billion. Because the waterway's overall revenues now exceed $2 billion per year, the investment appears more than worthwhile.

Yet Panamanian officials must now contend with problems inconceivable to Roosevelt and his successors. Global climate change is rendering it much more difficult to manage all kinds of infrastructural systems built during periods of relative climatic stability. Recent severe drought events have lowered Gatun and Alajuela Lakes, which supply fresh water both for the canal and for Panama's growing population, enough to force limits on cargo ships and thereby forfeit millions of dollars of tolls. For the canal to remain viable, officials must now consider building expensive, disruptive new networks of reservoirs, dams, and tunnels for storing and transporting fresh water.[43] As anthropogenic carbon emissions alter climate patterns, the assumption that technocrats can easily manage unintended ecological consequences is no longer tenable. Knowledge about likely bioenvironmental effects is now essential, not merely desirable, for making sound infrastructural investments.

LARGE-SCALE PROJECTS REQUIRE the convergence of many forces—political, economic, technological, scientific, and environmental—to take shape. It is also the case that grand infrastructural visions of the future sometimes fail to crystallize despite powerful coalitions in favor of them. Checking the technocratic impulse to solve complex problems with environmentally disruptive technological solutions requires political will, analytical rigor, and awareness of the options foreclosed by high-modernist plans for accelerating modernity.

Throughout its many phases, the Central American sea-level canal inspired visions of development that held both liberating and constricting implications for the anticipated host countries. The seaway proposals of the 1960s and 1970s served different U.S. presidential goals for improving relations with Panama as the original waterway and the Canal Zone slid into technopolitical obsolescence.

The proposals also had important, unexpected repercussions for environmental management and associated concepts of progress. Environmental scientists and activists did not cause the cancellation of the seaway in any of its iterations. They did, however, use the proposal to open up new discussions about the harmful consequences of maritime-induced bioinvasions, and about the kinds of science needed to quantify and predict the negative effects of marine invasive species on ecological and economic systems. They also influenced later generations; most relevant to current events are the biologists who put pressure on President

Johnson's Canal Study Commission, setting an example that informed scientific responses five decades later to the proposed Nicaragua Grand Canal.[44]

Unbuilt projects merit attention for many reasons, including the ways in which the planned and improvisational work underlying them influenced decision-making at the time and in ensuing eras. The resurgence of the sea-level canal proposal at strategic points in the intertwined history of the United States and Panama provides windows into moments of diplomatic, technological, scientific, and environmentalist transformation. Such historical moments in turn remind us of the value of envisioning alternative futures, and of questioning technocratic prescriptions that promise to modernize landscapes and societies without ensuring environmental quality and equal justice for all.

NOTES

Introduction. The Central American Sea-Level Canal and the Environmental History of Unbuilt Megaprojects

1. See, e.g., Caumartin, Review of *Emperors in the Jungle.*

2. David Kirsch, "Project Plowshare"; Frenkel, "A Hot Idea"; Scott Kirsch, *Proving Grounds*; Kaufman, *Project Plowshare.*

3. Lindsay-Poland, *Emperors in the Jungle.*

4. APICSC, *Interoceanic Canal Studies 1970*, 41.

5. Pritchard, "Joining Environmental History with Science and Technology Studies."

6. George Collins, "Introduction," 12.

7. Collins, 7. The book was Ponten, *Architektur.*

8. Goldin and Lubell, *Never Built Los Angeles*; Goldin and Lubell, *Never Built New York*; Ovnick, "*Never Built Los Angeles*"; Will Heinrich, "Remember When They Wanted to Build a Parking Lot over the Hudson?," *New York Times*, Sept. 21, 2017.

9. Carse and Kneas, "Unbuilt and Unfinished."

10. Oberdeck, "Archives of the Unbuilt Environment." See also Hindle, "Levees That Might Have Been"; Hindle, "Prototyping the Mississippi Delta."

11. Heffernan, "Bringing the Desert to Bloom"; Heffernan, "Shifting Sands," 618.

12. Lehmann, "Infinite Power to Change the World," 99. See also Fleming, *Fixing the Sky.*

13. Scott, *Seeing Like a State.* See also Loren Graham, *Ghost of the Executed Engineer.*

14. Macfarlane, "Negotiated High Modernism," 326. See also Reuss, "Seeing Like an Engineer."

15. Teller, "The Plowshare Program."

16. Regis, *Monsters*, 172. See also Fleming, *Fixing the Sky*, chap. 5.

17. Broderick, *Reconstructing Strangelove*, 69–72.

18. Alexis Madrigal, "7 (Crazy) Civilian Uses for Nuclear Bombs," *Wired*, Apr. 10, 2009; Dave Gilson and Adam Weinstein, "8 of the Wackiest (or Worst) Ideas for Nuclear Weapons," *Mother Jones*, Nov. 9, 2011; "Revealed: Madcap 1960s Plan to Use 23 Nuclear Bombs to Blast through California Mountains and Make Way for Highway," *Daily Mail*, Sept. 25, 2014; Ed Regis, "What Could Go Wrong? The Insane 1950s Plan to Use H-Bombs to Make Roads and Redirect Rivers," *Slate*, Sept. 30, 2015.

19. Kaufman, *Project Plowshare*, 2.

20. On the political role of scientific expertise, see Bocking, *Ecologists and Environmental Politics.*

21. O'Neill, *Firecracker Boys.*

22. David Kirsch, "Project Plowshare," 216; O'Neill, *Firecracker Boys*, 83; Scott Kirsch, *Proving Grounds*, 77.

23. Scott Kirsch and Mitchell, "Earth-Moving as the 'Measure of Man' "; Millar and Mitchell, "Spectacular Failure, Contested Success"; Rothschild, "Environmental Awareness in the Atomic Age"; Cittadino, "Paul Sears and Plowshare."

24. O'Neill, "Project Chariot," 36; Scott Kirsch, *Proving Grounds*, 105–6; Egan, *Barry Commoner and the Science of Survival.*

25. Reed, "Ecological Investigation in the Arctic," 372; Peter Coates, "Project Chariot"; Wilt with Hacker, "Gifts of a Fertile Mind."

26. Scott Kirsch, *Proving Grounds*, 76; Scott Kirsch and Mitchell, "Earth-Moving as the 'Measure of Man,' " 129.

27. Heffernan, "Bringing the Desert to Bloom," 108; Scott Kirsch, *Proving Grounds*, 8.

28. Kaufman, *Project Plowshare*, 222 (quote), 69; Scott Kirsch and Mitchell, "Earth-Moving as the 'Measure of Man,' " 129; Bocking, *Ecologists and Environmental Politics*, 90. See also Buys, "Isaiah's Prophecy"; Findlay, *Nuclear Dynamite*; Krygier, "Project Ketch."

29. Bent Flyvbjerg, "Mega Delusional: The Curse of the Megaproject," *New Scientist*, Nov. 30, 2013. See also Flyvbjerg, "Survival of the Unfittest."

30. Primack and Hippel, *Advice and Dissent*, 173; Luther Carter, "Rio Blanco."

31. See, e.g., Taylor, *Making Bureaucracies Think*; Caldwell, *National Environmental Policy Act*; Clark and Canter, *Environmental Policy and NEPA*; Lindstrom and Smith, *National Environmental Policy Act*; Dreyfus and Ingram, "National Environmental Policy Act"; Andrews, *Managing the Environment, Managing Ourselves.*

32. Rome, "What Really Matters in History?"; Sutter, "The World with Us."

33. Sutter, "Tropical Conquest and the Rise of the Environmental Management State."

34. See, e.g., Stine, *Mixing the Waters*; Espeland, *Struggle for Water.*

35. Lifset, *Power on the Hudson*; Noll and Tegeder, *Ditch of Dreams*; Shawn Miller, "Minding the Gap"; Ficek, "Imperial Routes"; Davis, *Everglades Providence*; Conway, *High-Speed Dreams*; Suisman, "American Environmental Movement's Lost Victory."

36. Rozwadowski, "Engineering, Imagination, and Industry."

37. Peyton, *Unbuilt Environments*, 8, 11.

38. Peyton, 18 (quote), 14. See also Scott, *Seeing Like a State*; Li, "Beyond 'the State' and Failed Schemes."

39. Kohler, *Landscapes and Labscapes*; Sutter, "Nature's Agents or Agents of Empire?"; Vetter, *Knowing Global Environments*; Bocking, "Situated but Mobile," Laura Martin, "Proving Grounds," Hersey and Vetter, "Shared Ground." See also Billick and Price, *Ecology of Place*; Scoville, "Hydraulic Society and a 'Stupid Little Fish.' "

40. Rankin, "Zombie Projects"; Rowe, "Promises, Promises"; d'Avignon, "Shelf Projects"; Carse and Kneas, "Unbuilt and Unfinished," 22.

41. Redfield, *Space in the Tropics*, 16; Carse and Kneas, "Unbuilt and Unfinished," 15–17.

42. Humboldt to Kelley, May 12, 1856.

43. Lyndon B. Johnson, "Remarks on the Decision to Build a Sea Level Canal and to Negotiate a New Treaty with Panama," in *PPPUS*, 1963–64, 2: 809.

44. Adas, *Dominance by Design*.

45. Covich, "Projects That Never Happened"; Covich, "Frank Golley's Perspectives."

46. On the blurring of boundaries between technology and nature, see, e.g., White, *Organic Machine*; Stine and Tarr, "At the Intersection of Histories"; Reuss and Cutcliffe, *Illusory Boundary*.

Chapter 1. Canalizing and Colonizing the Isthmus

1. Jaén Suárez, *Hombres y Ecología en Panamá*; Conniff, *Black Labor on a White Canal*; Frenkel, "Geography, Empire, and Environmental Determinism"; Castro Herrera, "On Cattle and Ships"; Lindsay-Poland, *Emperors in the Jungle*; Newton, *Silver Men*; Sutter, "Nature's Agents or Agents of Empire?"; Sutter, "Tropical Conquest and the Rise of the Environmental Management State"; Julie Greene, *Canal Builders*; Carse, *Beyond the Big Ditch*; Raby, "Ark and Archive"; Carse et al., "Panama Canal Forum"; Lasso, *Erased*.

2. Brady, "Historical Geography of the Earliest Colonial Routes"; McCullough, *Path between the Seas*; Delgado et al., *Maritime Landscape of the Isthmus of Panamá*.

3. "Humboldt," *New York Times*, Sept. 15, 1869, 1. English language monographs include Helferich, *Humboldt's Cosmos*; Sachs, *Humboldt Current*; Walls, *Passage to Cosmos*; Rebok, *Humboldt and Jefferson*; Wulf, *Invention of Nature*; Echenberg, *Humboldt's Mexico*.

4. Nathaniel Rich, "The Very Great Alexander von Humboldt," *New York Review of Books*, Oct. 22, 2015, 37.

5. Humboldt, *Political Essay*, 1:35, 4:22; Humboldt and Bonpland, *Personal Narrative*, 6:242–43.

6. Humboldt, *Personal Narrative*, 6:245.

7. Humboldt, *Political Essay*, 1:18. On the Raspadura Canal see Humboldt, *Personal Narrative*, 6:260; Frederick Collins, "The Isthmus of Darien," 149.

8. Humboldt, *Political Essay*, 1:27, 1:25.

9. See, e.g., Willis Johnson, *Four Centuries of the Panama Canal*.

10. Humboldt, *Personal Narrative*, 6:285.

11. Humboldt, 6:240, 6:248.

12. Humboldt, 6:245 (emphasis in original).

13. Humboldt, *Political Essay*, 1:35–36.

14. Humboldt, *Personal Narrative*, 6:281.

15. Humboldt, 6:288–89.

16. Humboldt, 6:276.

17. See, e.g., Humboldt, *Political Essay*, 2:24.

18. Anthony, "Mining as the Working World of Alexander von Humboldt's Plant Geography."

19. Humboldt, *Personal Narrative*, 6:297.

20. Humboldt, 6:297–98.

21. Niles, *History of South America and Mexico*, 2:6.

22. Lloyd, "Account of Levellings Carried across the Isthmus of Panama"; Humboldt, *Views of Nature*, 292.

23. Humboldt, *Views of Nature*, 292.

24. Murchison, *Address to the Royal Geographical Society of London*, 64; Bidwell, *Isthmus of Panamá*, 99; Balf, *Darkest Jungle*, 59.

25. Mack, *Land Divided*; Velásquez Runk, "Creating Wild Darien."

26. McGuinness, *Path of Empire*.

27. De Lesseps, "The Panama Canal"; McCullough, *Path between the Seas*; Clayton, "The Nicaragua Canal in the Nineteenth Century"; Brannstrom, "Almost a Canal."

28. Kelley, *Union of the Oceans by Ship-Canal without Locks*, 7; Fitz-Roy, "Considerations on the Great Isthmus of Central America."

29. Humboldt to Kelley, May 12, 1856; "Baron von Humboldt's Encouragement"; Buel, "Piercing the American Isthmus"; Parks, *Colombia and the United States*, 334–35.

30. Humboldt to Kelley, May 12, 1856 (emphasis in original). See also Kelley, "On the Connection between the Atlantic and Pacific Oceans."

31. Buel, "Piercing the American Isthmus," 276; Walker, *Report of the Isthmian Canal Commission*, 72; Mehaffey, *Isthmian Canal Studies—1947*, 32.

32. de Lesseps, "The Panama Canal"; McCullough, *Path between the Seas*.

33. Frederick Collins, "The Isthmus of Darien and the Valley of the Atrato," 148, 161–62.

34. Walker, *Report of the Isthmian Canal Commission*, 56.

35. McCullough, *Path between the Seas*.

36. Mahan, *Influence of Sea Power*, 83.

37. Mahan, 34–35.

38. Roland, Bolster, and Keyssar, *Way of the Ship*, chap. 31; Smith, *Boundless Sea*.

39. Pérez, *War of 1898*.

40. Mahan, "The Panama Canal and Sea Power in the Pacific," 155.

41. McCullough, *Path between the Seas*.

42. For English language scholarly studies, see, e.g., Ealy, *Yanqui Politics and the Isthmian Canal*; McCullough, *Path between the Seas*; Hogan, *Panama Canal in American Politics*; Parker, *Panama Fever*; Maurer and Yu, *Big Ditch*.

43. McCullough, *Path between the Seas*; Sutter, "Nature's Agents or Agents of Empire?"

44. *Report of the Board of Consulting Engineers for the Panama Canal*, 12.

45. *Report of the Board of Consulting Engineers for the Panama Canal*, 35, 61.

46. *Report of the Board of Consulting Engineers for the Panama Canal*, 99, 100.

47. For more detail on the mechanics of the isthmian sea-level canal's requisite tidal lock, as articulated by a later generation of engineers, see Mehaffey, *Isthmian Canal Studies—1947*, 66, 73, 75, 96.

48. *Report of the Board of Consulting Engineers for the Panama Canal*, 82, 85, 91.

49. "President Theodore Roosevelt Message to the Congress, February 19, 1906," in *ICPQ*, 448–49.

50. Haskin, *Panama Canal*, 13–14.

51. Sutter, "Nature's Agents or Agents of Empire?"

52. Nida, *Panama and Its "Bridge of Water."*

53. Lasso, *Erased*; Carse, *Beyond the Big Ditch*, chap. 7; Carse, " 'Like a Work of Nature.' "

54. Michael Donoghue, "The Panama Canal and the United States"; Missal, *Seaway to the Future.*

55. Allen, *Our Canal in Panama*, 22–23.

56. Missal, *Seaway to the Future*; Henderson, "The Face of Empire"; Strong, "Jimmy Carter and the Panama Canal Treaties," 271.

57. LaFeber, *Panama Canal*; Maurer and Yu, *Big Ditch.*

58. Mahan, "The Panama Canal and Sea Power in the Pacific," 166.

59. Roosevelt, "On American Motherhood," 262.

60. Mahan, "The Panama Canal and Sea Power in the Pacific," 178.

61. Humboldt, *Cosmos*, 1:368.

62. English language analyses of Humboldt's imperialist legacy include Pratt, *Imperial Eyes*; Sachs, "The Ultimate 'Other' "; Rupke, "A Geography of Enlightenment"; Walls, *Passage to Cosmos.*

63. Schwarz, "Alexander von Humboldt's Visit to Washington"; Walls, *Passage to Cosmos*, 121.

64. Cushman, "Humboldtian Science," 22.

65. Frenkel, "Geography, Empire and Environmental Determinism"; Carse, *Beyond the Big Ditch*, 72–73.

66. Bennett, *History of the Panama Canal*, 146.

67. Niles, *History of South America and Mexico*, 2:6.

68. Nida, *Panama and Its "Bridge of Water,"* 84.

Chapter 2. Confronting the Canal's Obsolescence

1. *Report of the Board of Consulting Engineers for the Panama Canal*, 10.

2. Conniff, *Black Labor on a White Canal*; Newton, *Silver Men*; Julie Greene, *Canal Builders.*

3. Allen, *Our Canal in Panama*, 30.

4. McCullough, *Path between the Seas*, 613.

5. Heckadon-Moreno, "Light and Shadows," 33; Carse, *Beyond the Big Ditch*, 115–16; Mehaffey, *Isthmian Canal Studies—1947*, 76.

6. "Public Resolution—No. 99—70th Congress," in Rea and Shield, *Statements for the Seventieth Congress*, 478–79.

7. Panama Canal Company, *The Panama Canal: The Third Locks Project*, 1.

8. Mehaffey, *Isthmian Canal Studies—1947*, 52, app. 3; Travis and Watkins, "Control of the Panama Canal," 410–11; Conn, Engelman, and Fairchild, *Guarding the United States and Its Outposts*, chap. 12.

9. Bowman, "Puzzle in Panama," 414.

10. Bowman, 407.

11. John G. Clayburn quoted in "Bomb-Proof Canal at Panama Sought," *New York Times*, Jan. 17, 1946; "Sea Level Canal at Panama Urged," *New York Times*, Jan. 22, 1948; James Reston, "Visit to Canal by Royall Linked to Atom Defenses," *New York Times*, Feb. 10, 1948.

12. Mehaffey, *Isthmian Canal Studies—1947*.

13. Mehaffey, 16, 62.

14. Mehaffey, 65, 85–86.

15. Mehaffey, 110, 97.

16. *FRUS, 1955–1957*, 7:129; Travis and Watkins, "Control of the Panama Canal," 410–11. See also *FRUS, 1948*, 9:336.

17. Flood, "Panama Canal Questions: Immediate Action Required," *U.S. Congressional Record*, May 8, 1963, in *ICPQ*, 230.

18. LaFeber, Panama Canal; Maurer and Yu, *Big Ditch*.

19. Travis and Watkins, "Control of the Panama Canal," 417.

20. Knapp, *Red, White, and Blue Paradise*; Missal, *Seaway to the Future*; Moore, *Empire on Display*; Michael Donoghue, *Borderland on the Isthmus*.

21. Travis and Watkins, "Control of the Panama Canal," 416; Conniff, *Panama and the United States*, 79–81.

22. *FRUS, 1955–1957*, 7:152.

23. *FRUS, 1955–1957*, 7:157.

24. *FRUS, 1955–1957*, 7:154.

25. *Report of the Board of Consulting Engineers for the Panama Canal*, 10; APICSC, *Interoceanic Canal Studies 1970*, V-23.

26. *FRUS, 1955–1957*, 7:177.

27. Michael Donoghue, *Borderland on the Isthmus*, 176, 252; Lindsay-Poland, "U.S. Military Bases in Latin America and the Caribbean."

28. Travis and Watkins, "Control of the Panama Canal," 416.

29. "Republic of Panama Flag Planted in Canal Zone by Students in Surprise Move," *Star & Herald* (Panama), May 3, 1958, in *ICPQ*, 31–33; Ralph K. Skinner, "Students Harass Government—Riots Neutralize Panama Gains," *Christian Science Monitor*, May 31, 1958, in *ICPQ*, 47–50. See also Tate, "The Panama Canal and Political Partnership"; LaFeber, *Panama Canal*, 124–131.

30. Tate, "The Panama Canal and Political Partnership," 128; LaFeber, *Panama Canal*, 129; Conniff, *Panama and the United States*, 81–83; National Declassification Center, U.S. National Archives and Records Administration, "The Panama Canal: Riots, Treaties, Elections, and a Little Military Madness, 1959–1973" (2015), https://www.archives.gov/research/foreign-policy/panama-canal.

31. Flood, "July 23, 1958: Panama Canal: Object of Irresponsible Political Extortion," in *ICPQ*, 51–72; Flood, "Monroe Doctrine or Khrushchev Doctrine?," *Congressional Record*, Apr. 12, 1962, in *ICPQ*, 153; Flood, "Panama Canal: Key Target of Fourth Front," *Congressional Record*, Apr. 19, 1960, in *ICPQ*, 135.

32. Hanson W. Baldwin, "The Panama Canal—II," *New York Times*, Aug. 13, 1960.

33. Laleh Khalili, "How the (Closure of the) Suez Canal Changed the World," *The Gamming* (blog), Aug. 31, 2014, https://thegamming.org/2014/08/31/how-the-closure-of-the-suez-canal-changed-the-world/.

34. Teller et al., *Constructive Uses of Nuclear Explosives*, vi. For more on this origin story, see Scott Kirsch, *Proving Grounds*, 11; Kaufman, *Project Plowshare*, 13–14.

35. Dwight D. Eisenhower, "Atoms for Peace Speech," Dec. 8, 1953, on the International Atomic Energy Agency's website, https://www.iaea.org/about/history/atoms-for-peace-speech. See also, e.g., Weart, *Rise of Nuclear Fear*, chap. 8.

36. Graves, *Engineer Memoirs*, 85–86.

37. Edward Teller, "We're Going to Work Miracles," *Popular Mechanics*, Mar. 1960, 97–101, 278, 280, 282 (quote on 100).

38. Hays, *Conservation and the Gospel of Efficiency*.

39. Gerald W. Johnson to John O. Pastore, Nov. 27, 1963, NSF, Files of Charles E. Johnson, Box 36, Folder Nuclear—Nuclear Excavation (Sea Level Canal), LBJL.

40. Reines, "The Peaceful Nuclear Explosion"; Vortman, "Excavation of a Sea-Level Ship Canal," 88.

41. Teller, "We're Going to Work Miracles," 98.

42. O'Neill, "Project Chariot"; O'Neill, *Firecracker Boys*; David Kirsch, "Project Plowshare"; Scott Kirsch and Mitchell, "Earth-Moving as the 'Measure of Man' "; Millar and Mitchell, "Spectacular Failure, Contested Success"; Scott Kirsch, *Proving Grounds*; Kaufman, *Project Plowshare*; Cittadino, "Paul Sears and Plowshare."

43. Teller, "We're Going to Work Miracles," 99.

44. O'Neill, *Firecracker Boys*.

45. Wolfe, "The Ecological Aspects of Project Chariot," 62; Golley, *History of the Ecosystem Concept in Ecology*, 74.

46. Leopold, *A Sand County Almanac*.

47. Wolfe, "The Ecological Aspects of Project Chariot," 65–66.

48. *Isthmian Canal Plans—1960, Annex VII*, Box 75, Papers of William Merrill Whitman, DDEL; "Transcript of 2nd Meeting," Entry A1 36040-C, Container 1, RG 220, NARA.

49. Isthmian Canal Studies Board of Consultants, *Report to the Committee on Merchant Marine and Fisheries*, 7. See also "Transcript of 2nd Meeting," Entry A1 36040-C,

Container 1, RG 220, NARA; APICSC, *Interoceanic Canal Studies 1970*, V-15; W. W. Whitman to Bonner, May 5, 1964, NSF, Files of Charles E. Johnson, Box 36, Folder Nuclear—Nuclear Excavation (Sea Level Canal), LBJL.

50. Flood, "Isthmian Canal Policy—An Evaluation," *Congressional Record*, June 7, 1962, in *ICPQ*, 177.

51. *FRUS, 1961–1963*, 12:392.

52. Chiari to Kennedy, Sept. 8, 1961, repr. in *Star & Herald* (Panama), Nov. 16, 1961, in *ICPQ*, 162. On State Department efforts to discourage Chiari from sending the letter, see *FRUS, 1961–1963*, 12:394.

53. Kennedy to Chiari, Nov. 2, 1961, in *ICPQ*, 164–65; *FRUS, 1961–1963*, 12:400.

54. *FRUS, 1961–1963*, 12:401. See also Graves, *Engineer Memoirs*.

55. *FRUS, 1961–1963*, 12:401. Around this time, Ball, the under secretary of state, also started urging Kennedy to avoid U.S. involvement in Vietnam, and he later helped President Carter rally senatorial support for the ratification of the 1977 Panama Canal Treaties. Robert D. McFadden, "George W. Ball Dies at 84; Vietnam's Devil's Advocate," *New York Times*, May 28, 1994.

56. *FRUS, 1961–1963*, 12:402.

57. *FRUS, 1961–1963*, 12:400.

58. *FRUS, 1961–1963*, 12:403. On Chiari's grievances, see *FRUS, 1961–1963*, 12:405; *FRUS, 1961–1963*, 12:408.

59. "Joint Communique and Aide Memoire Resulting from Discussions in Panama between United States and Panamanian Representatives," Jan. 8, 1963, in *ICPQ*, 220–21.

60. O'Neill, "Project Chariot"; O'Neill, *Firecracker Boys*; David Kirsch, "Project Plowshare"; Scott Kirsch and Mitchell, "Earth-Moving as the 'Measure of Man' "; Millar and Mitchell, "Spectacular Failure, Contested Success"; Scott Kirsch, *Proving Grounds*; Kaufman, *Project Plowshare*; Rodgers, "From a Boon to a Threat."

61. "Hydrogen Explosion Set Off Underground in Nevada," *New York Times*, July 7, 1962; "Giant H-Bomb Shot Rips Crater in Nevada Desert," *Los Angeles Times*, July 7, 1962. See also Kelly, "Moving Earth and Rock with a Nuclear Device"; Kelly testimony in U.S. Congress, *Second Transisthmian Canal*, 51; Nevada National Security Site, *Sedan Crater* (Naval Nuclear Security Administration, 2013), https://www.nnss.gov/docs/fact_sheets/DOENV_712.pdf.

62. Kaufman, *Project Plowshare*, 86–87; Hacker, *Fallout from Plowshare*, 7–8; "Suit Filed in Leukemia Deaths," *New York Times*, June 16, 1984.

63. See, e.g., "Atomic Earth Mover," *Newsweek*, July 16, 1962; "When Nuclear Bomb Is Harnessed for Peace," *U.S. News & World Report*, Dec. 10, 1962; "Digging with H-Bombs," *Business Week*, May 18, 1963; "An Atomic Blast to Help Build a U.S. Canal?," *U.S. News & World Report*, May 20, 1963; "Another 'Panama Canal': A-Blasts May Do the Job," *U.S. News & World Report*, June 10, 1963.

64. Flood, *ICPQ*, 453–57, 348–50.

65. Flood, "Panama Canal Questions: Immediate Action Required," *U.S. Congressional Record*, May 8, 1963, in *ICPQ*, 230–31; Flood, "Focus of Power Politics," *U.S. Congressional Record*, Mar. 9, 1964 in *ICPQ*, 304–18.

66. See, e.g., Maass, *Muddy Waters*; Shallat, *Structures in the Stream*.

67. See, e.g., Egan, *Barry Commoner and the Science of Survival*; Benjamin Greene, *Eisenhower, Science Advice, and the Nuclear Test-Ban Debate*.

68. U.S. Congress, *Nuclear Test-Ban Treaty*, 210, 265.

69. Langer, "Project Plowshare"; "Nuclear Ditch-Digging," *Business Week*, Dec. 21, 1963; "Nuclear Energy: Ploughshare Canals," *Time*, Jan. 31, 1964.

70. Department of State Memorandum of Conversation to Merrill Whitman, Gerald W. Johnson, V. Lansing Collins, and H. Franklin Irwin, Nov. 22, 1963, NSF, Files of Charles E. Johnson, Box 36, Folder Nuclear—Nuclear Excavation (Sea Level Canal), LBJL.

71. Gerald W. Johnson to John O. Pastore, Nov. 27, 1963, NSF, Files of Charles E. Johnson, Box 36, Folder Nuclear—Nuclear Excavation (Sea Level Canal), LBJL.

Chapter 3. Mobilizing for Panama Canal II

1. Department of State Memorandum of Conversation to Merrill Whitman, Gerald W. Johnson, V. Lansing Collins, and H. Franklin Irwin, Nov. 22, 1963, NSF, Files of Charles E. Johnson, Box 36, Folder Nuclear—Nuclear Excavation (Sea Level Canal), LBJL.

2. Lawrence, "Exception to the Rule?," 40, 45.

3. Adas, *Dominance by Design*.

4. "Canal Called Not Vital to Navy, but Zone Is a U.S. Military Hub," *New York Times*, Jan. 11, 1964.

5. Doel and Harper, "Prometheus Unleashed." See also Harper, *Make It Rain*, chap. 7.

6. See, e.g., Adas, *Dominance by Design*, chap. 6; McNamara, *In Retrospect*; Ford, *CIA and the Vietnam Policymakers*.

7. Zierler, *Invention of Ecocide*; Martini, *Agent Orange*.

8. See, e.g., Doel and Harper, "Prometheus Unleashed"; Doel, "Scientists as Policymakers"; Dorsey, "Dealing with the Dinosaur"; Hamblin, *Oceanographers and the Cold War*; McNeill and Unger, *Environmental Histories of the Cold War*; Hecht, *Entangled Geographies*; Hamblin, *Arming Mother Nature*; Audra Wolfe, *Competing with the Soviets*; Bocking and Heidt, *Cold Science*.

9. Flood, "Congress Must Save the Panama Canal," *Congressional Record*, Apr. 9, 1963, in *ICPQ*, 211; "More Panama Flags to Go Up in Canal Zone," *Panama American*, Oct. 28, 1963, in *ICPQ*, 291–92.

10. Rowland Evans and Robert Novak, "Ugly Americans," *Washington Post*, Nov. 1, 1963, in *ICPQ*, 301–2.

11. " 'Firm' Policy on Panama Favored by Americans," *Washington Post*, Feb. 12, 1964.

12. *FRUS, 1964–1968*, 31:372; Jorden, *Panama Odyssey*; LaFeber, *Panama Canal*, 138–40; McPherson, "From 'Punks' to Geopoliticians"; McPherson, *Yankee, No!*, chap. 3.

13. "Let's Act Our Age in Panama," *Life*, Jan. 24, 1964, 4. See also Trevor Armbrister, "Panama: Why They Hate Us," *Saturday Evening Post* 237, Mar. 7, 1964, 75–79.

14. Michael Donoghue, *Borderland on the Isthmus*, 247.

15. "Gunfire Flares: Relations Severed till Pacts Are Altered Chiari Asserts," *New York Times*, Jan. 11, 1964; "Canal Called Not Vital to Navy, but Zone Is a U.S. Military Hub," *New York Times*, Jan. 11, 1964; "Job Differences Persist in Zone Despite U.S. Equal-Pay Policy," *New York Times*, Jan. 12, 1964.

16. Belinfante, Petren, and Vakil, *Report on the Events in Panama, January 9–12, 1964*.

17. McPherson, "Courts of World Opinion."

18. Kaufman, *Project Plowshare*, 99–100.

19. The U.S. government owned the Panama Canal Company and was represented by the secretary of the army, whom the annual reports referred to as the stockholder and personal representative of the U.S. president. See, e.g., Panama Canal Company and Canal Zone Government, *Annual Report: 1965*, 1.

20. Vance, Memorandum for the President, Feb. 10, 1964, NSF, Box 65, Folder 2, LBJL.

21. "Text of the Second Half of 1964 Republican Platform," *New York Times*, July 13, 1964.

22. *FRUS, 1964–1968*, 11:7; Bundy, Memorandum for Holders of NSAM No. 282, July 20, 1964, Files of Charles E. Johnson, Box 41, Folder 1, LBJL; Hacker, *Fallout from Plowshare*, 9; Kaufman, *Project Plowshare*, 109–10. On the broader ways in which natural forces influenced the U.S. nuclear testing program, see Oatsvall, "Weather, Otters, and Bombs."

23. U.S. Congress, *Second Transisthmian Canal*, 12–13.

24. Lyndon B. Johnson, "Remarks Following the Signing of a Joint Declaration with Panama," Apr. 3, 1964, in *PPPUS*, 1963–64, 1:245.

25. *FRUS, 1964–1968*, 31:408; Thomas Mann interview by Jorden, p. 24, Box 22, Personal Papers of William J. Jorden, LBJL.

26. Lawrence, "Exception to the Rule?"

27. Robert Anderson interview by Jorden, May 14, 1979, pp. 1, 4, Box 21, Personal Papers of William J. Jorden, LBJL. See also Jorden, *Panama Odyssey*, chap. 5; Ashley Morrow, "An Unexpected Journey: Spacecraft Transit the Panama Canal," NASA, Apr. 9, 2015, https://www.nasa.gov/content/goddard/an-unexpected-journey-spacecraft-transit-the-panama-canal.

28. Panama Review Group Meeting [Minutes], Apr. 7, 1964, NSF, Box 66, Folder 4, LBJL.

29. Panama Review Group Meeting [Minutes], Apr. 7, 1964.

30. Sandars, *America's Overseas Garrisons*; Lutz, *Bases of Empire*.

31. Belinfante, Petren, and Vakil, *Report on the Events in Panama, January 9–12, 1964*.

32. Merrill Whitman, Panama Review Group Meeting [Minutes], Apr. 7, 1964, NSF, Box 66, Folder 4, LBJL.

33. *FRUS, 1964–1968*, 31:420. See also "Briefing Paper: Panama," Apr. 28, 1964, NSF, Box 66, Folder 3, LBJL.

34. Stephen Ailes, Memorandum to the President on Canal Zone Policies, July 23, 1964, NSF, Box 66, Folder 6, LBJL.

35. *FRUS, 1964–1968*, 31:417.

36. *FRUS, 1964–1968*, 31:421. See also *FRUS, 1964–1968*, 31:420.

37. On this argument, see "Briefing Paper: Panama," Apr. 28, 1964, NSF, Box 66, Folder 3, LBJL; *FRUS, 1964–1968*, 31:420.

38. *FRUS, 1964–1968*, 31:417.

39. *FRUS, 1964–1968*, 31:417.

40. See, e.g., Ted Szulc, "Crisis in Panama Spurs U.S. Study of a New Canal," *New York Times*, Jan. 20, 1964; John W. Finney, "Cost of Atom-Dug Sea-Level Canal Is Put at $500 Million," *New York Times*, Jan. 21, 1964.

41. Graves et al., *Isthmian Canal Studies—1964*, 1.

42. Graves et al., *Isthmian Canal Studies—1964*, 3, 42.

43. U.S. Congress, *Authorizing the President to Appoint a Commission*, 2.

44. "Determination of Site for Construction of a Sea Level Canal Connecting the Atlantic and Pacific Oceans," *Congressional Record—Senate*, Mar. 26, 1964, 6467.

45. U.S. Congress, *Second Transisthmian Canal*, 30. See also Lawrence Galton, "A New Canal—Dug by Atom Bombs," *New York Times*, Sept. 20, 1964.

46. U.S. Congress, *Second Transisthmian Canal*, 34. See also Fleming, *Fixing the Sky*, 201–4; Adler, *Neptune's Laboratory*, 119.

47. Frederick G. Dutton to Herbert C. Bonner, Apr. 22, 1964, NSF, Files of Charles E. Johnson, Box 36, Folder Nuclear—Nuclear Excavation (Sea Level Canal), LBJL; Edward A. McDermott to Herbert C. Bonner, May 5, 1964, NSF, Files of Charles E. Johnson, Box 36, Folder Nuclear—Nuclear Excavation (Sea Level Canal), LBJL; "Interoceanic Canal Problem: Inquiry or Cover Up? Sequel," *Congressional Record*, July 29, 1965, in *ICPQ*, 453–516; "Statement of the Honorable Daniel J. Flood . . . Before the Committee on Merchant Marine and Fisheries, House of Representatives," June 4, 1964, in *ICPQ*, 457–60.

48. "Interoceanic Canal Problem," in *ICPQ*, 453–57; Jorden, *Panama Odyssey*, 99; Spear, *Daniel J. Flood*, 71.

49. Robert M. Sayre to McGeorge Bundy, Dec. 4, 1964, Folder 6, Box 67, Country File: Latin America—Panama, NSF, Papers of Lyndon B. Johnson, LBJL.

50. McPherson, *Yankee No!*, 112–14.

51. "Statement of Panama Review Committee, November 6, 1964," NSC Histories, Panama Crisis, 1964, Folder 1, LBJL.

52. *FRUS, 1964–1968*, 31:419.

53. *FRUS, 1964–1968*, 31:420.

54. *FRUS, 1964–1968*, 31:420.

55. Even Robert Anderson referred to the 1903 treaty as "an emotional problem with Panama." *FRUS, 1964–1968*, 31:421. See also McPherson, "Rioting for Dignity."

56. *FRUS, 1964–1968*, 31:420. See also *FRUS, 1964–1968*, 31:421, 31:423.

57. Harry McPherson interview by Jorden, Mar. 28, 1979, Box 22, Personal Papers of William J. Jorden, LBJL.

58. Lyndon B. Johnson, "Remarks on the Decision to Build a Sea Level Canal and to Negotiate a New Treaty with Panama," Dec. 18, 1964, in *PPPUS*, 1963–64, 2:809; Ted

Szulc, "U.S. Decides to Dig a New Canal at Sea Level in Latin America and Renegotiate Panama Pact," *New York Times*, Dec. 19, 1964.

59. Lawrence, "Exception to the Rule?"

60. Paul P. Kennedy, "Panama Ponders Policy on Canal," *New York Times*, Dec. 21, 1964.

61. *FRUS, 1964–1968*, 31:422.

62. Paul P. Kennedy, "Panama Observes Riot Anniversary," *New York Times*, Jan. 10, 1965.

63. Remarks by Chairman Robert B. Anderson at the APICSC meeting with Secretaries of State, Treasury, Commerce, Army; Deputy Secretary of Defense; Chairman of AEC; Chairman of Joint Chiefs of Staff; President's Assistant for Science and Technology, Mar. 17, 1966, Office Files of Harry McPherson, Box 12, Folder Panama Canal, LBJL.

64. Graves, *Engineer Memoirs*, 89.

65. McNeill and Unger, "Introduction: The Big Picture," in *Environmental Histories of the Cold War*, 16.

Chapter 4. Navigating High Modernism

1. Daniel J. Flood, "Preconceived Plan for Sea-Level Canal Destroyed: Time for Action on Terminal Lake–Third Locks Plan Has Come," *Congressional Record*, May 20, 1970, H4619–H4622; Strom Thurmond, "Introduction of the Panama Canal Modernization Act," *Congressional Record—Senate*, Feb. 10, 1971, 2453–60; Commoner, *Closing Circle*, 60; Boffey, "Sea-Level Canal"; Newman, "National Academy of Science Committee"; Dunson, "Sea Snakes and the Sea Level Canal Controversy"; Rubinoff, "Sea-Level Canal in Panama"; Beeton, *Report of the Committee on Ecological Effects*; Sapp, *What Is Natural*, 125; Sapp, *Coexistence*, 103.

2. Lyndon B. Johnson, "Message to the Congress Transmitting First Annual Report of the Atlantic-Pacific Interoceanic Canal Study Commission," August 3, 1965, in *PPPUS*, 1965, 1:400.

3. See especially Scott Kirsch, *Proving Grounds*, 5–6.

4. Raymond Hill, "28th Meeting (9–10 July 1970) Transcript," p. 224, Container 5, RG 220, NARA.

5. Loo, "People in the Way."

6. Scott, *Seeing Like a State*.

7. Walt Rostow to Dean Rusk, Jan. 25, 1965, Folder 2, Box 67, Country File: Latin America—Panama, NSF, Papers of Lyndon B. Johnson, LBJL.

8. Carson, *Silent Spring*, 127.

9. Macfarlane, "Negotiated High Modernism."

10. See, e.g., Rycroft and Szyliowicz, "Decision-Making in a Technological Environment"; Josephson, *Industrialized Nature*; Zipp, *Manhattan Projects*; Zierler, *Invention*

of Ecocide; Martini, *Agent Orange*; McNeill and Unger, *Environmental Histories of the Cold War*.

11. See also Frenkel, "A Hot Idea?"; Lindsay-Poland, *Emperors in the Jungle*, chap. 3.

12. Audra Wolfe quoted in Daniel A. Gross, "Can a Nuclear Explosion Be Peaceful? US Scientists Used to Think So," *Public Radio International*, July 11, 2017, https://www.pri.org/stories/2017-07-11/can-nuclear-explosion-be-peaceful-us-scientists-used-think-so.

13. An Act to Provide for an Investigation and Study to Determine a Site for the Construction of a Sea Level Canal connecting the Atlantic and Pacific Oceans, Public Law 88–609, Sept. 22, 1964, https://www.gpo.gov/fdsys/pkg/STATUTE-78/pdf/STATUTE-78-Pg990.pdf.

14. Langer, "ACDA"; Foster, "Risks of Nuclear Proliferation."

15. *FRUS, 1964–1968*, 31:424.

16. "Interoceanic Canal Problem: Inquiry or Cover Up?," *Congressional Record*, Apr. 1, 1965, in *ICPQ*, 432, 433.

17. McPherson, "Courts of World Opinion," 92; Daniel J. Flood to Stephen Ailes, Jan. 25, 1965, in *ICPQ*, 444–46; Spear, *Daniel J. Flood*, 70. Flood's own subsequent volume of his addresses on the Panama Canal from 1958 to 1966—*ICPQ*—totaled 523 pages. See also "Bibliography of Panama Canal Issues," *Congressional Record—Senate*, July 10, 1967, 18114–19.

18. Sheffey to Harry C. McPherson Jr., Aug. 25, 1965, Box 4, Folder Personnel thru 1965, APICSC Administrative File, Entry 36040-F, RG 220, NARA.

19. Sheffey, "Transcript of 11th Mtg—23 June 1966," p. 19, Container 2, RG 220, NARA.

20. Noble, "Transcript of Proceedings—14th Mtg—22 June 67," p. 114, Container 2, RG 220, NARA.

21. Jorden, *Panama Odyssey*, 107; Sheffey interview by Jorden, May 8, 1979, Box 23, Personal Papers of William J. Jorden, LBJL.

22. Eisenhower, *Wine Is Bitter*.

23. Jorden, *Panama Odyssey*, 107; Phillips, Hall, and Black, *Reining in the Rio Grande*, chap. 7; "A Guide to the Raymond A. Hill Papers, 1890–1945," Briscoe Center for American History, accessed Mar. 29, 2020, https://legacy.lib.utexas.edu/taro/utcah/01461/cah-01461.html.

24. Lawrence Van Gelder, "K.E. Fields, Engineer Who Led Atomic Energy Unit, Dies at 87," *New York Times*, July 10, 1996; Alice Buck, *The Atomic Energy Commission* (U.S. Department of Energy, 1983), https://www.energy.gov/sites/prod/files/AEC%20History.pdf.

25. Edward Tomlinson, "More Panama Demands Likely—Ike's Brother Stirs Trouble," *Washington Daily News*, July 18, 1958; Flood, "Panama Canal and the Milton Eisenhower Paper," *Congressional Record*, May 21, 1964, in *ICPQ*, 399–408; "Interoceanic Canal Problem: Inquiry or Cover Up? Sequel," *Congressional Record*, July 29, 1965, in *ICPQ*, 455.

26. Sheffey did not assume the official title of executive director until mid-1967; prior

to then, he was called the acting executive director or executive secretary, but his duties remained the same. Sheffey to Anderson, Aug. 11, 1965, Box 4, Folder Personnel thru 1965, APICSC Administrative File, Entry 36040-F, RG 220, NARA; "Transcript of Proceedings—14th Mtg—22 June 67," Container 2, RG 220, NARA; "Col. John Sheffey, 70, Dies; Expert on the Panama Canal," *Washington Post*, Nov. 24, 1989.

27. Sheffey interview by Jorden, May 8, 1979, p. 22, Box 23, Personal Papers of William J. Jorden, LBJL.

28. Scott Kirsch, *Proving Grounds*, 161–63; Graves, *Engineer Memoirs*.

29. Noble replaced Woodbury in June 1967, and Groves replaced Noble in June 1969. "Transcript of Proceedings—14th Mtg—22 June 67," Container 2, RG 220, NARA; "Transcript of Proceedings—Twenty-Third Meeting (24 June 1969)," Container 4, RG 220, NARA.

30. Graves, "Nuclear Excavation of a Sea-level, Isthmian Canal," 369.

31. U.S. Congress, *Peaceful Applications of Nuclear Explosives*, 2–53.

32. "Transcript of Second Meeting," June 11, 1965, Container 1, RG 220, NARA.

33. "Transcript of Second Meeting," June 11, 1965; "Transcript of 5th Meeting—17–18 Sept 65," Container 1, RG 220, NARA.

34. UPI, "Johnson Urges Speed Up of $$ for Canal Survey," *El Panama America*, Aug. 4, 1965; "Funds for Canal Surveys," *Philadelphia Inquirer*, Aug. 8, 1965.

35. "Transcript of Second Meeting," June 11, 1965, Container 1, RG 220, NARA.

36. "Transcript of Third Meeting," July 16, 1965, Container 1, RG 220, NARA.

37. See, e.g., Hagen, *Entangled Bank*; Golley, *History of the Ecosystem Concept in Ecology*; Bocking, *Ecologists and Environmental Politics*, Laura Martin, "Proving Grounds."

38. Jorden, *Panama Odyssey*, 97–98; Irwin, *Interview with John N. Irwin II*, by Gordon W. Evans, May 30, 1991; Nick Ravo, "John N. Irwin II, 86, Diplomat and Ex-Aide to MacArthur," *New York Times*, Feb. 29, 2000.

39. New York Zoological Society, *1965 Annual Report*, 89–90.

40. "Transcript of Third Meeting," July 16, 1965, Container 1, RG 220, NARA.

41. Rubinoff, "Mixing Oceans and Species," 69. Recent studies date the emergence of the Panama land bridge and sea barrier to around 2.8 million years ago, but during the 1960s researchers estimated that the eastern Pacific and western Atlantic marine populations had remained separated for as little as 2 million years and as much as 5 million years or more. See, e.g., Woodring, "The Panama Land Bridge as a Sea Barrier," 425; Rubinoff, "Central American Sea-Level Canal," 858; O'Dea et al., "Formation of the Isthmus of Panama."

42. "Transcript of Third Meeting," July 16, 1965, Container 1, RG 220, NARA.

43. "Transcript of Third Meeting," July 16, 1965.

44. "Transcript of 5th Meeting—17–18 Sept 65," Container 1, RG 220, NARA; "Transcript of 8th Meeting—16 Dec 65," Container 2, RG 220, NARA.

45. See, e.g., Robertson, *Malthusian Moment*; Flippen, *Conservative Conservationist*.

46. "Transcript of 5th Meeting—17–18 Sept 65," p. 64, Container 1, RG 220, NARA.

47. "Transcript of 5th Meeting—17–18 Sept 65," p. 135.

48. Willard Libby quoted in "Plowshare Future Tied to Soviets," *Christian Science Monitor*, Apr. 6, 1966.

49. Weart, *Rise of Nuclear Fear*. See also Scott Kirsch, *Proving Grounds*, 38–39.

50. "Transcript of Third Meeting," July 16, 1965, Container 1, RG 220, NARA.

51. APICSC, *Interoceanic Canal Studies 1970*, V-23.

52. "Press Release: US Canal Study Commission Inspecting Routes in Panama," Aug. 21, 1965, APICSC Scrapbook, Container 1, RG 220, NARA.

53. "Transcript—7th Meeting (18 Nov 65)," Container 2, RG 220, NARA.

54. "Transcript—7th Meeting (18 Nov 65)."

55. Lyndon B. Johnson, "Annual Budget Message to the Congress, Fiscal Year 1967," Jan. 24, 1966, in *PPPUS*, 1966, 1:26.

56. Seaborg, "Transcript of 9th Meeting—14 January 1966," Container 2, RG 220, NARA.

57. Hill, "Transcript of 9th Meeting—14 January 1966," Container 2, RG 220, NARA.

58. Kaufman, *Project Plowshare*, 129–33.

59. "Transcript of 8th Meeting—16 Dec 65," p. 55, Container 2, RG 220, NARA.

60. "Inter-Sea Canal Aired," *Miami Herald*, May 20, 1966; "The Panama Canal Study Hurting for Time, Money," *Oakland Times*, May 26, 1966. See also "LBJ Told Poverty Dented, But Canal Plans Slowed," *Tulsa Daily World*, Aug 16, 1966.

61. "Panamanians Marking Riots Burn a U.S. Flag," *New York Times*, Jan. 10, 1966.

62. Richard E. Bevir, "We Built Our Own Road through the Darien Gap," *Popular Mechanics*, Aug. 1961, 206. See also Joseph C. Ingraham, "American Roller-Coaster," *New York Times*, Feb. 7, 1960; "2 in Heavy-Duty Car Blaze Trail through Panama's Dense Jungle," *New York Times*, June 13, 1960; Kip Ross, "We Drove Panama's Darién Gap," *National Geographic* 119 (1961): 368–89; Hanbury-Tenison and Burton, "Should the Darien Gap Be Closed?"

63. These figures came from the 1960 census, as reported in Torres de Araúz, *Demographic Characteristics of Human Groups*. See also Anthony Coates, *Central America*.

64. "Army Engineer Unit Set Up in Zone for New Canal Study," *New York Times*, Aug. 8, 1965; A. G. Sutton, "Sea-Level Canal," *Military Engineer* 60 (1968): 105.

65. Reuss, "The Art of Scientific Precision."

66. Reeve Waring, "Darien Sea-Level Canal Survey Starts," *Sunday American* (Panama), Mar. 20, 1966; "Darien Indians Protest US 'Invasion,'" *Panama American*, Mar. 22, 1966. See also "The Panama Canal Study Hurting for Time, Money," *Oakland Times*, May 26, 1966.

67. Technical Liaison Officer and Staff, Press Analysis, June 1, 1966, APICSC Scrapbook (quote), RG 220, NARA; "AEC Studying Nuclear Effects in R.P. Minister Confirms Atomic Canal Study," *Sunday American* (Panama), Apr. 10, 1966; "Bar Flays Eleta-Adair Notes on Sea-Level Canal Studies," *Panama American*, Jan. 18, 1967.

68. "Panama tab: 'Confidential: APICSC Public Information Study Group Questionnaire for Panama,'" n.d. [ca. Feb. 21, 1967], Folder Public Information Questionnaires and Responses (Confidential), Container 12, RG 220, NARA. The information about Eleta's

degree is from "Fallece Fernando Eleta Almarán, pionero de la televisión panameña," Telemetro.com, Aug. 12, 2011, http://www.telemetro.com/nacionales/Fallece-Fernando-Eleta-Almaran-television_0_395660436.html. See also Lindsay-Poland, *Emperors in the Jungle*, 93, 231n61.

69. OICS Technical Liaison Officer and Staff, Press Analysis, June 1, 1966, APICSC Scrapbook, RG 220, NARA; Kenneth Smart, "Panama's Big Ditch Poses Big Problems," *Dallas Times Herald*, Apr. 10, 1966; Bruce Biossat, "If Panama Canal Is on Way Out, Then What?," *Washington Daily News*, Apr. 11, 1966; John M. Goshko, "Planned A-Tests Alarm Latins," *Washington Post*, Apr. 21, 1966; "Latin Nations Differ on Atom-Free Pact," *Washington Post*, Apr. 21, 1966; John W. Finney, "U.S. Delays A-Test as Aid to Treaty," *New York Times*, Feb. 11, 1967; "Sucre, US Envoy Play Hosts at 'Atoms in Action' Opening," *Panama American*, May 8, 1967; Lindsay-Poland, *Emperors in the Jungle*, 94, 231n62.

70. Christobal Sarmiento, "What's Happening in Darien Survey?," *Panama American*, May 29, 1966.

71. OICS Technical Liaison Officer and Staff, Press Analysis, June 1, 1966, APICSC Scrapbook, RG 220, NARA; "Transcript of 11th Mtg—23 June 1966," p. 33, Container 2, RG 220, NARA.

72. Reuters, "Bulldozers Prowl for New Canal Site," *Christian Science Monitor*, June 6, 1966.

73. Howe, *People Who Would Not Kneel*; Lindsay-Poland, *Emperors in the Jungle*, 95–96, 101.

74. "Transcript of 11th Mtg—23 June 1966," p. 22, Container 2, RG 220, NARA.

75. Bevir, "We Built Our Own Road through the Darien Gap," *Popular Mechanics*, Aug. 1961; Howe, *Chiefs, Scribes, and Ethnographers*, 194; Scott Doggett, "Birds versus Buzz Saws in Jungle Joust," *Los Angeles Times*, Sept. 21, 2004.

76. Christobal Sarmiento, "What's Happening in Darien Survey?," *Panama American*, May 29, 1966 (quote); "Transcript of 11th Mtg—23 June 1966," p. 37, Container 2, RG 220, NARA. See also "Jungle Quest for Answers to Future Canal Near End," *Star & Herald* (Panama), Feb. 14, 1968.

77. The sea-level canal studies by Reina Torres de Araúz include *Human Ecology Studies, Panama, Phase 1: Final Report* (Columbus, Ohio: Battelle Memorial Institute, 1967); *Demographic Characteristics of Human Groups Inhabiting the Eastern Region of the Republic of Panama* (Columbus, Ohio: Battelle Memorial Institute, 1968); *Shrimp Fishery of Panama* (Columbus, Ohio: Battelle Memorial Institute, 1968); "Demographic and Dietary Data for Human Groups Inhabiting the Eastern Region of the Republic of Darien," *BioScience* 19 (1969): 331–35; and *Human Ecology of Route 17 (Sasardi-Morti) Region, Darien, Panama*, transl. and ed. F. W. McBryde (Columbus, Ohio: Battelle Memorial Institute, 1970).

78. Battelle Memorial Institute, *Environmental Impact Assessment for Darien Gap Highway*.

79. "Panama Presses New Canal Study: Scientists Assay Nuclear Excavation in the

Jungles," *New York Times*, Feb. 27, 1967. See also Judy Burton, "Storey Inspects Proposed Sites for New Canal," *Dallas Morning News*, Apr. 20, 1967.

80. "US Pushes Study of Route for Sea-Level Canal in RP," *Weekend American* (Panama), Feb. 25, 1967; Colin Hale, "Canal Studies in Darien Are Not All Milk and Honey," *Panama American*, Apr. 3, 1967.

81. APICSC, *Third Annual Report*, July 31, 1967, Entry A1 36040-D, Container 3, RG 220, NARA.

82. Henry Giniger, " 'Martyrs' of Riot Hailed in Panama," *New York Times*, Jan. 10, 1967.

83. "Canal Studies Agreements Identical in RP, Colombia," *Panama American*, Mar. 2, 1967; Panama Canal Information Office, Daily Digest of News and Editorial Opinion of Panama News Media, n.d. [ca. Mar. 4, 1967], APICSC Scrapbook, RG 220, NARA.

84. "Agitators Try to Stir Canal Studies Trouble," *Panama American*, June 12, 1967.

85. Arlen J. Large, "Blocked Canal," *Wall Street Journal*, Sept. 6, 1967; "US Sea-Level Canal Studies Going Smoothly, OICS Says," *Panama American*, Sept. 7, 1967; "Sea-Level Canal Studies Going Well, Chief Says," *Star & Herald* (Panama), Sept. 8, 1967; "OICS Phasing Out Studies of RP Sea-Level Canal Site," *Panama American*, Sept. 8, 1967.

86. AEC, "AEC to Conduct Plowshare Cratering Experiment in Nevada," Jan. 23, 1967, APICSC Scrapbook, RG 220, NARA; Representative Craig Hosmer quoted in "AEC May Disrupt New Canal Plans: Congressman Denounces Act of Atom Group," *Star & Herald* (Panama), Feb. 11, 1967.

87. "Fallout Spread over Six States from Nevada Test, AEC Reports," *Washington Post*, Apr. 29, 1966; John W. Finney, "U.S. Delays A-Test as Aid to Treaty," *New York Times*, Feb. 11, 1967; Lindsay-Poland, *Emperors in the Jungle*, 99; Hunt, "Mexican Nuclear Diplomacy."

88. Victor K. McElheny, "U.S. Peaceful A-Use Program Opposed," *Washington Post*, Mar. 13, 1967.

89. John W. Finney, "Maneuvering Is On over 'Panama Canal II,' " *New York Times*, Nov. 27, 1966.

90. "Panama–United States Joint Announcement of Agreement on Texts of Three New Canal Treaties, June 26, 1947," in Congressional Research Service, *Background Documents Relating to the Panama Canal*, 1148; "Surrender in Panama," *Chicago Tribune*, June 28, 1967; Chesly Manly, "Full Panama Treaty Text," *Chicago Tribune*, July 15, 1967; LaFeber, *Panama Canal*, 147–48; Major, *Prize Possession*.

91. "Transcript of Proceedings—16th Mtg—7 Sept 67," Container 2, RG 220, NARA; Dean Rusk, Department of State Bulletin, Aug. 7, 1967, in Congressional Research Service, *Background Documents*, 1148, note 78.

92. Arlen J. Large, "Blocked Canal," *Wall Street Journal*, Sept. 6, 1967.

93. David Kirsch, "Project Plowshare"; Scott Kirsch, *Proving Grounds*.

94. Morgenstern and Heiss, *General Report on the Economics of the Peaceful Uses of Underground Nuclear Explosions*, 5 (emphasis in original)

95. Martini, *Agent Orange*; Zierler, *Invention of Ecocide*.

96. David Kirsch, *Electric Car and the Burden of History*; Laird, *Solar Energy, Technology Policy, and Institutional Values*.

97. Inglis and Sandler, "Special Report on Plowshare," 49.

98. "Transcript of Proceedings—16th Mtg—7 Sept 67," p. 10, Container 2, RG 220, NARA.

Chapter 5. Assessing Mankind's Most Gigantic Biological Experiment

1. Bowler, *Earth Encompassed*, 208.

2. See, e.g., *BioScience* 19 (1969); William Martin et al., *Symposium on Sea-level Canal Bioenvironmental Studies*; Torres de Araúz, *Demographic Characteristics of Human Groups*; APICSC, *Annual Reports*, Entry A1 36040-D, Container 3, RG 220, NARA.

3. See, e.g., Covich, "Projects That Never Happened"; Covich, "Frank Golley's Perspectives"; John Motyka, "James Duke, 88, Globe-Trotting Authority on Healing Plants, Is Dead," *New York Times*, Dec. 5, 2018; G. Wayne Clough, personal communication; Torres de Araúz, *Demographic Characteristics of Human Groups*.

4. Kuhn, *Structure of Scientific Revolutions*.

5. Quote from "Transcript of Proceedings—18th Mtg—7 March 1968," Entry A1 36040-C, Container 4, RG 220, NARA.

6. Rubinoff, "Central American Sea-Level Canal," 857.

7. See, e.g., O'Neill, *Firecracker Boys*.

8. Seaborg, *Stemming the Tide*, 326–39; Kaufman, *Project Plowshare*, 129–33.

9. "Canals and the Atom," *Evening Star* (Washington, D.C.), Apr. 18, 1968.

10. Thomas O'Toole, "Super-Clean Nuclear Explosive Now Appears within U.S. Grasp," *Washington Post*, May 12, 1968.

11. "RP Envoy Scores Nuclear Canal Idea," *Panama American*, n.d. [ca. Aug. 1968].

12. John W. Finney, "Maneuvering Is On over 'Panama Canal II,' " *New York Times*, Nov. 27, 1966; Bruce Biossat, "If Panama Canal Is on Way Out, Then What?," *Washington Daily News*, Apr. 11, 1966; "Gulf Oil Orders Six 300,000-Ton Japanese Ships," *Washington Star*, May 9, 1966.

13. Tania Long, "Lloyd's Register Plans 500,000-Ton Tanker," *New York Times*, Jan. 4, 1966; "1 million-Ton Cargo Ship Called Feasible," *Washington News*, Dec. 14, 1966.

14. "Transcript of Proceedings—18th Mtg—7 March 1968," p. 98, Entry A1 36040-C, Container 4, RG 220, NARA. See also Graves, *Engineer Memoirs*; Coulombe, "Searching for Stability."

15. "Transcript of Proceedings—18th Mtg—7 March 1968," p. 98, Entry A1 36040-C, Container 4, RG 220, NARA.

16. "Transcript of the Executive Session of the 26th Meeting of the APICSC, 12 March 1970," Entry A1 36040-C, Container 5, RG 220, NARA.

17. Milton S. Eisenhower to John Sheffey, Mar. 11, 1968, Box 1: Foreign Policy Study 1965–68, Folder 1: Foreign Policy Study, 1965 thru 1968, Entry 36040-H, RG 220, NARA.

18. "Transcript of Proceedings—16th Mtg—7 Sept 67," pp. 17–18, Entry A1 36040-C, Container 2, RG 220, NARA.

19. See, e.g., Tate and Allen, "The Proposed New Treaties for the Panama Canal," 276.

20. "Transcript of Proceedings—19th Mtg—23 May 1968," Entry A1 36040-C, Container 4, RG 220, NARA.

21. Lyndon B. Johnson, "Message to the Congress Transmitting the Fourth Annual Report of the Atlantic-Pacific Interoceanic Canal Study Commission," Sept. 5, 1968, in *PPPUS*, 1968–1969, 2:466.

22. Leonard Carmichael to Glenn Seaborg, Mar. 14, 1963, RU 155, Box 57, Folder General Correspondence—1963–1965, SIA; Carmichael to James E. Reeves, Aug. 13, 1963, RU 155, Box 57, Folder General Correspondence—1963–1965, SIA.

23. Wallen, "A Long Term Biological Study of the Marine Organisms on Both Sides of the Middle American Isthmus," 1965, RU 108, Box 41, Folder 6, SIA; John N. Wolfe to I. E. Wallen, July 19, 1965, RU 108, Box 42, Folder 2, SIA; Harve J. Carlson to I. E. Wallen, Sept. 21, 1965, RU 108, Box 42, Folder 2, SIA; Charles L. Dunham to T. Dale Stewart, Jan. 4, 1966, RU 155, Box 59, Folder A Long Term Biological Study of Marine Organisms, SIA.

24. Cowan et al., "Meetings"; Ritterbush, "Biology and the Smithsonian Institution"; S. Dillon Ripley to James E. Reeves, June 2, 1964, RU 155, Box 57, Folder General Correspondence—1963–1965, SIA; I. E. Wallen to Dr. Ripley, Mar. 13, 1964, RU 155, Box 57, Folder General Correspondence—1963–1965, SIA.

25. I. E. Wallen to S. Dillon Ripley, Dec. 14, 1964, RU 155, Box 57, Folder General Correspondence—1963–1965, SIA; Wallen, "Proposal to Atomic Energy Commission," Dec. 21, 1964, RU 108, Box 42, Folder 2, SIA.

26. Golley et al., "The Structure of Tropical Forests in Panama and Colombia"; Ethan Walther to I. E. Wallen, Jan. 8, 1964, RU 108, Box 42, Folder 2, SIA.

27. Wallen to Ripley, Feb. 25, 1965, RU 108, Box 42, Folder 2, SIA; Wallen to Ripley, Apr. 16, 1965, RU 108, Box 42, Folder 2, SIA.

28. John N. Wolfe to I. E. Wallen, July 19, 1965, RU 108, Box 42, Folder 2, SIA.

29. Ripley, *Trail of the Money Bird*, 14–15.

30. Henson, "Baseline Environmental Survey."

31. Raby, "Ark and Archive"; Raby, *American Tropics*.

32. Richards, "What the Tropics Can Contribute"; Ripley, "Perspectives in Tropical Biology"; AIBS Committee, "Preliminary Report: Panama Tropical Center for Biological Research," 1965, RU 108, Box 42, Folder 2, SIA; Wallen to Ripley, June 7, 1965, RU 108, Box 42, Folder 2, SIA.

33. Hagen, "Problems in the Institutionalization of Tropical Biology."

34. Woodring, "Panama Land Bridge as a Sea Barrier."

35. Henson, "Baseline Environmental Survey."

36. Mayr, "Geographic Speciation in Tropical Echinoids," 2; Mayr to Ira Rubinoff, Feb. 28, 1966, Box 13, Folder 900, Papers of Ernst Mayr, HUA. Courtesy of the Harvard University Archives.

37. Mayr to Harvey Brooks, Jan. 8, 1965, Box 11, Folder 862, Papers of Ernst Mayr, HUA.

38. Mayr to Harvey Brooks, Jan. 8, 1965.

39. Milam, "Equally Wonderful Field."

40. Haffer, *Ornithology, Evolution, and Philosophy*, 130–31.

41. Rubinoff to Mayr, Apr. 22, 1961, Box 8, Folder 795, Papers of Ernst Mayr, HUA; Rubinoff to Mayr, May 23, 1962, Box 8, Folder 795, Papers of Ernst Mayr, HUA.

42. Rubinoff, Statement, *Hearings before the Subcommittee on the Panama Canal*; Christen, "At Home in the Field."

43. Ripley, "Announcement," Apr. 18, 1966, RU 254, Box 39, Folder 1, SIA.

44. Rubinoff, "Mixing Oceans and Species."

45. Rubinoff, "Mixing Oceans and Species," 72; Aronova, Baker, and Oreskes, "Big Science and Big Data in Biology."

46. Rubinoff to Mayr, Nov. 26, 1965, Box 13, Folder 900, Papers of Ernst Mayr, HUA; Rosenblatt, "Zoogeographic Relationships of the Marine Shore Fishes"; Briggs, "Relationship of the Tropical Shelf Regions."

47. R. S. Davidson to Sidney R. Galler, May 6, 1966, RU 155, Box 57, Folder General Correspondence—1966, SIA. See also William E. Martin to Sidney R. Galler, Aug. 10, 1966, RU 155, Box 58, Folder General Correspondence—1966, SIA.

48. Galler to Ripley, May 16, 1966, RU 136, Box 1, Folder 3, SIA.

49. R. S. Cowan to [MNH] Department Chairmen, Dr. Moynihan and Dr. Klein, Mar. 14, 1966, RU 155, Box 57, Folder General Correspondence—1966, SIA; Cowan to Ripley, Apr. 20, 1966, RU 155, Box 58, Folder General Correspondence—1966, SIA.

50. H. G. Woodbury, Memorandum for the Record, Aug. 8, 1966, RU 155, Box 58, Folder General Correspondence—1966, SIA.

51. "Detailed Minutes: Meeting on Tropical Research," Aug. 3, 1966, RU 155, Box 58, Folder General Correspondence—1966, SIA; William E. Martin, "BMI Report on Official Travel," Aug. 8, 1966, RU 155, Box 58, Folder General Correspondence—1966, SIA.

52. John P. Sheffey to Sidney R. Galler, Aug. 29, 1966, RU 108, Box 3, Folder 5, SIA.

53. Cowan to Ripley, through Galler, Aug. 26, 1966, RU 155, Box 58, Folder General Correspondence—1966, SIA.

54. "Panama Conference on Tropical Biology," RU 155, Box 58, Folder Panama Conference Transcript, SIA; Buechner and Fosberg, "Contribution toward a World Program."

55. Galler, Memorandum for the Files, Feb. 8, 1968, RU 108, Box 3, Folder 5, SIA.

56. Rubinoff and Rubinoff, "Interoceanic Colonization of a Marine Goby"; Rubinoff and Rubinoff, "Observations on the Migration of a Marine Goby."

57. These authors included Chesher, Topp, Menzies, Glynn, Dunson, Newman, and Jones, and to a lesser extent, Graham, Hubbs, Kropach, Porter, and Vermeij. "Bibliography of Research Supported through the Facilities of the STRI Marine Laboratories during their First Ten Years," 1977, RU 329, Box 125, Folder 3, SIA.

58. Rubinoff, "Mixing Oceans and Species."

59. Aron and Por, "Tribute to Heinz Steinitz (1909–1971) and Gunnar Thorson (1906–1971)."

60. Por, "One Hundred Years of Suez Canal"; Safriel, "The 'Lessepsian Invasion.'"

61. Woodbury to Ripley, Nov. 25, 1966; Ripley to Woodbury, Dec. 13, 1966, RU 108, Box 3, Folder 5, SIA; "Panama Conference on Tropical Biology," p. 30, RU 155, Box 58, Folder Panama Conference Transcript, SIA.

62. Ripley to Woodbury, Jan. 23, 1967, RU 108, Box 3, Folder 5, SIA; Charles C. Noble to S. Dillon Ripley, Sept. 14, 1967, RU 99, Box 50, Folder Sea Level Canal, SIA; Galler to Ripley, Dec. 7, 1967, RU 108, Box 3, Folder 5, SIA; F. Raymond Fosberg, "Program for Ecological Research in Relation to the Proposed Sea-Level Canal," Dec. 1, 1967, RU 136, Box 1, Folder 3, SIA.

63. Galler to Ripley, Mar. 10, 1967, RU 470, Box 37, Folder NSF Paper, SIA.

64. Galler, Memorandum for the Files, Feb. 8, 1968, RU 108, Box 3, Folder 5, SIA; Rubinoff to Russell B. Stevens, Aug. 28, 1969, Folder Biology & Agriculture: CERIC: Data Accumulation, 1969, NAS.

65. Cole, "Can the World Be Saved?," *BioScience*; LaMont C. Cole, "Can the World Be Saved?," *New York Times Magazine*, Mar. 31, 1968, 35, 95, 97, 100.

66. Rubinoff, "Central American Sea-Level Canal," 857.

67. See also Hubbs and Pope, "Spread of the Sea Lamprey through the Great Lakes"; Aron and Smith, "Ship Canals and Aquatic Ecosystems."

68. Rubinoff, "Central American Sea-Level Canal," 858.

69. Rubinoff, 860.

70. "Text of Kennedy's Address to Academy of Natural Sciences," *New York Times*, Oct. 23, 1963, 24; Rubinoff, "Central American Sea-Level Canal," 860.

71. Rubinoff, "Central American Sea-Level Canal," 861.

72. Dreyfus and Ingram, "National Environmental Policy Act"; Taylor, *Making Bureaucracies Think*; Caldwell, *National Environmental Policy Act*; Andrews, *Managing the Environment, Managing Ourselves*; Lindstrom and Smith, *National Environmental Policy Act*.

73. Moon et al., "Multidisciplinary Conceptualization of Conservation Opportunity," 1488.

74. Rubinoff to Russell B. Stevens, Aug. 28, 1969, Folder Biology & Agriculture: CERIC: Data Accumulation, NAS.

75. See, e.g., Kingsland, *Evolution of American Ecology*; Kristin Johnson, "Natural History as Stamp Collecting"; Hagen, "Problems in the Institutionalization of Tropical Biology"; Billick and Price, *Ecology of Place*; Raby, "Ark and Archive."

76. See especially Norse and Crowder, *Marine Conservation Biology*, xviii; Kroll, *America's Ocean Wilderness*; Oreskes, "Scaling Up Our Vision"; Rozwadowski, *Vast Expanses*.

77. Charles Elton, *Ecology of Invasions by Animals and Plants*, 100.

78. Szylvian, "Transforming Lake Michigan"; Mann, *Exotic Species in Mariculture*.

79. Kinsey, "'Seeding the Water as the Earth'"; Keiner, *Oyster Question*.

80. Elton, *Ecology of Invasions*, 96. See also Carlton, "Blue Immigrants"; Nelson, "Ravages of Teredo."

81. Elton, *Ecology of Invasions*, 94; Hildebrand, "Panama Canal as a Passageway for Fishes"

82. Rubinoff and Rubinoff, "Interoceanic Colonization of a Marine Goby," 477.

83. Sheffey, "When Caribbean and Pacific Waters Mix," 1329. See also Sheffey to Philip H. Abelson, Oct. 22, 1968, RU 108, Box 3, Folder 5, SIA.

84. Rubinoff, letter to editor, 762.

85. John Hillaby, "A Risky Mix," *New Scientist*, 1969, 280–81.

86. Briggs, "Panama's Sea-Level Canal"; Briggs, *Marine Zoogeography*.

87. Briggs, "Panama's Sea-Level Canal," 512.

88. Briggs, "The Sea-Level Panama Canal," 47.

89. Rubinoff, "Sea-Level Canal Controversy"; Porter, "Ecology and Species Diversity of Coral Reefs."

90. Voss, "Biological Results of the University of Miami Deep-Sea Expeditions," 56.

91. Carl Hubbs to Philip Handler, Feb. 17, 1969, Folder Biology & Agriculture: CERIC 1969, NAS.

92. See, e.g., Doel, "Constituting the Postwar Earth Sciences."

93. Hubbs, "Need for Thorough Inventory," 468; Aron and Smith, "Ship Canals and Aquatic Ecosystems."

94. Hubbs, "Need for Thorough Inventory," 467.

95. Topp, "Interoceanic Sea-Level Canal," 1324. STRI scientists confirmed this hypothesis decades later, though without referencing Topp. Smith, Bell, and Bermingham, "Cross-Cordillera Exchange."

96. Menzies, "Transport of Marine Life."

97. Chesher, "Transport of Marine Plankton."

98. Topp, "Interoceanic Sea-Level Canal," 1326.

99. Sheffey to Burton Benjamin, Apr. 7, 1969, RU 108, Box 42, Folder 1, SIA.

100. Rubinoff, "Sea-Level Canal Controversy," 34; Meredith Jones, "Panamic Biota"; Dawson, "Occurrence of an Exotic Eleotrid Fish"; Jones and Dawson, "Salinity-Temperature Profiles in the Panama Canal Locks." See also Carlton, "Transoceanic and Interoceanic Dispersal of Coastal Marine Organisms," 319.

101. APICSC, "Fifth Annual Report," July 31, 1969, pp. 3–7, Entry A1 36040-D, Container 3, RG 220, NARA; Richard Nixon, "Message to the Congress Transmitting Annual Report of the Atlantic-Pacific Interoceanic Canal Study Commission," Aug. 6, 1969, *PPPUS*, 1969, 314.

102. Thomas O'Toole, "U.S. A-Test Linked to Canada Fallout," *Washington Post*, Jan. 9, 1969; Kaufman, *Project Plowshare*, 169–70.

103. "RP Establishes Office to Study Canal Routes," *Panama American*, n.d. [ca. Nov 1968]; APICSC, "Fifth Annual Report," July 31, 1969, 9.

104. William H. Gorishek, "Dream of Nuclear Canal Turns into Pick and Shovel Nightmare," *Panama American*, Apr. 22, 1969.

Chapter 6. Avoiding an Elastic Collision with Knowledge

1. See, e.g., Gottlieb, *Forcing the Spring*; Egan, *Barry Commoner and the Science of Survival*; Rubinson, *Rethinking the American Antinuclear Movement*; Spears, *Rethinking the American Environmental Movement.*

2. Richard Nixon, "Statement about the National Environmental Policy Act of 1969," Jan. 1, 1970, in *PPPUS*, 1970, 2.

3. Flippen, *Nixon and the Environment.*

4. Boffey, "Sea-Level Canal"; Newman, "National Academy of Science Committee," 256 (quote); Leslie A. Pray, "Ernst Mayr Dies," *The Scientist*, Feb. 4, 2005; Sapp, *What Is Natural*; Sapp, *Coexistence.*

5. Ira Rubinoff to Ernst Mayr, Apr. 4, 1968, Folder Biology & Agriculture: CERIC: General, 1968, NAS; Carl Hubbs to Philip Handler, Feb. 17, 1969, Folder Biology & Agriculture: CERIC 1969, NAS.

6. Luther Carter, "National Academy of Sciences"; Boffey, *Brain Bank of America.*

7. LaMont C. Cole to Ernst Mayr, Feb. 13, 1968, Box 15, Folder 952, Papers of Ernst Mayr, HUA.

8. Mayr to Cole, Feb. 8, 1968, Box 15, Folder 952, Papers of Ernst Mayr, HUA.

9. Cole to Mayr, Feb. 13, 1968, Box 15, Folder 952 Mayr Papers, HUA. Cole's statement is poignant given that three years later the academy's governing council apparently cut him from a list of nominees due to his environmental activism. See Walsh, "National Academy of Sciences"; Rome, *Genius of Earth Day.*

10. Mayr to Frederick Seitz, Apr. 16, 1968, Folder Biology & Agriculture: CERIC: General, 1968, NAS.

11. Seitz to Mayr, Apr. 19, 1968, Folder Biology & Agriculture: CERIC: General, 1968, NAS; Seitz to Mayr, May 22, 1968 [quote]; Mayr to Frederick Seitz, Apr. 16, 1968, Folder BIOLOGY & Agriculture: CERIC: General, 1968, NAS.

12. Seitz to Ripley, May 12, 1969, RU 108, Box 42, Folder 1, SIA; Mayr to Ripley, May 16, 1969, RU 108, Box 42, Folder 1, SIA. For more detail, see Keiner, "Two-Ocean Bouillabaisse," 859–60.

13. Robert B. Anderson to S. Dillon Ripley, Aug. 9, 1968, RU 108, Box 41, Folder 8, SIA; David Challinor to S. D. Ripley, S. R. Galler, I. E. Wallen, L. M. Talbot, M. Moynihan, H. Buechner, T. Jorling, Nov. 13, 1968, RU 108, Box 41, Folder 8, SIA; Challinor to Ripley, Galler, Wallen, Talbot, Moynihan, Buechner, Jorling, Nov. 22, 1968, RU 108, Box 41, Folder 8, SIA.

14. Ripley to Anderson, Dec. 2, 1968, RU 108, Box 41, Folder 8, SIA; Paula Ullmann to Lynne Mac Elroy, Dec. 2, 1968, RU 99, Box 216, Folder Sea Level Canal, SIA.

15. Anderson to Seitz, Dec. 6, 1968, Folder Biology & Agriculture: CERIC: General, 1968, NAS.

16. Sheffey to C. E. Sunderlin, Dec. 30, 1968, Folder Biology & Agriculture: CERIC: Background Data, Battelle Memorial Inst Proposals, NAS; Russell B. Stevens to A. G. Norman, Jan. 13, 1969, Folder Biology & Agriculture: Interoceanic Canal, Com. on Ecological Research for: Battelle Inst. Report & Proposal, 1968–69, NAS; A. G. Norman to Stephen H. Spurr, Feb. 9, 1969, Folder Biology & Agriculture: CERIC: Beginning of Program, 1968–1969, NAS.

17. National Research Council, Division of Biology and Agriculture, "For Action New Projects," Feb. 9, 1969, Folder Biology & Agriculture: CERIC 1969, NAS.

18. B. L. Kropp and R. B. Stevens, "Proposal for Planning for Long-Term Ecological Studies of an Atlantic-Pacific Interoceanic Canal to Atlantic-Pacific Interoceanic Canal Study Commission, February 15, 1969–June 30, 1970," Folder Biology & Agriculture: CERIC 1969, NAS.

19. Mueller, "New Canal," 167.

20. Sheffey to Philip H. Abelson, Jan. 21, 1969, Folder Biology & Agriculture: CERIC 1969, NAS.

21. "Transcript—Proceedings of 21st Mtg (13 Jan 69)," Entry A1 36040-C, Container 4, RG 220, NARA.

22. "Transcript—Proceedings of 21st Mtg (13 Jan 69)."

23. "Transcript—Proceedings of 21st Mtg (13 Jan 69)."

24. "Transcript of Proceedings—Twenty-Third Meeting (24 June 1969)," pp. 76, 78, Entry A1 36040-C, Container 4, RG 220, NARA.

25. "Transcript of 24th Meeting (Oct 23, 1969)," Entry A1 36040-C, Container 4, RG 220, NARA.

26. *FRUS, 1969–1976*, E-10:524, E-10:521.

27. "Transcript of 24th Meeting (Oct 23, 1969)," p. 36, Entry A1 36040-C, Container 4, RG 220, NARA.

28. Henry Ramont, "2 Latin Lands Set New Canal Route," *New York Times*, Sept. 29, 1969; "RP, Colombia Propose New Canal to Cross Border of 2 Countries," *Panama American*, Sept. 30, 1969.

29. Nixon, "Statement about the National Environmental Policy Act of 1969."

30. Thompson, "Improving the Quality of Life," 2. For the 1959 hearings see U.S. Congress, *Biological and Environmental Effects of Nuclear War.*

31. Thompson, "Improving the Quality of Life," 4. On earlier environmental criticisms of the AEC, see, e.g., Balogh, *Chain Reaction*, 258–65; Hamblin, *Poison in the Well*, chap. 8.

32. Werth, "Closing Remarks," 1775.

33. Ramey, "Calvert Cliffs Campaign," 140. See also Balogh, *Chain Reaction*, chap. 8; Caldwell, *National Environmental Policy Act*, 43; Lindstrom and Smith, *National Environmental Policy Act*, 115–17.

34. Murphy, "National Environmental Policy Act and the Licensing Process," 963.

35. "Transcript of the Executive Session of the 26th Meeting of the APICSC, 12 March 1970," p. 20, Entry A1 36040-C, Container 5, RG 220, NARA.

36. "28th Meeting (9–10 July 1970) Transcript," pp. 85–86, Entry A1 36040-C, Container 5, RG 220, NARA.

37. Shor, Rosenblatt, and Isaacs, "Carl Leavitt Hubbs," 230.

38. "Transcript of the Executive Session of the 26th Meeting of the APICSC, 12 March 1970," p. 65, Entry A1 36040-C, Container #5, RG 220, NARA.

39. "Transcript of the Executive Session of the 26th Meeting of the APICSC, 12 March 1970," p. 65.

40. Joshua Lederberg, "Sea-Level Canal Points up Need for Environmental Data," *Washington Post*, Feb. 1, 1969.

41. Seitz to Mayr, Dec. 24, 1968, Folder Biology & Agriculture: CERIC: General, 1968, NAS; Mayr to Seitz, Jan. 6, 1969, Folder Biology & Agriculture: CERIC 1969, NAS.

42. Norman to R. B. Stevens, Mar. 3, 1969, Folder Biology & Agriculture: CERIC: Beginning of Program, 1968–1969, NAS; "Draft Minutes—Ad Hoc Group Meeting to Discuss Ecological Research Related to Sea-Level Canal, March 17, 1969," Folder Biology & Agriculture: CERIC: Beginning of Program, 1968–1969, NAS.

43. Mayr to Russell B. Stevens, Apr. 8, 1969, Box 17, Folder 1023, Papers of Ernst Mayr, HUA.

44. Mayr to Russell B. Stevens, May 19, 1969, Box 17, Folder 1023, Papers of Ernst Mayr, HUA; Stevens, "Memorandum to Participants, Meeting of March 17th," Apr. 14, 1969, Folder Biology & Agriculture: CERIC: Beginning of Program, 1968–1969, NAS; Alan R. Longhurst to Russell B. Stevens, Dec. 16, 1969, Folder Biology & Agriculture: CERIC: Subcom on Oceanography, Ad hoc, 1969, NAS; Stevens to Luna B. Leopold, Nov. 13, 1969, Folder Ad Hoc Committee on Hydrology, 1969–1970, NAS. For more detail, see Keiner, "Two-Ocean Bouillabaisse," 863–64.

45. Bayer, Voss, and Robins, *Bioenvironmental and Radiological Safety Feasibility Studies*, 9; Voss, "Biological Results of the University of Miami Deep-Sea Expeditions."

46. U.S. Congress, *Sea-Level Canal Studies*, 204–5.

47. MacArthur and Wilson, *Theory of Island Biogeography*.

48. Baker and Stebbins, *Genetics of Colonizing Species*; Blu Buhs, *Fire Ant Wars*.

49. National Research Council, Division of Biology and Agriculture, "Revised Summary Report—Workshop on Biology of Dispersal, Woods Hole, Mass., 20 Aug 1969," Folder Biology & Agriculture: Interoceanic Canal, Com. on Ecological Res for: Workshop on Biology of Dispersal, Aug. 20, 1969, NAS.

50. National Research Council, "Revised Summary Report."

51. Bakus to Russell B. Stevens, Apr. 30, 1970, Folder Biology & Agriculture: Interoceanic Canal, Com. on Ecological Res. for: Corres. re submission of Rept. & Comments thereon, 1970–73, NAS.

52. Bakus to Stevens, Dec. 2, 1969, Folder Biology & Agriculture: CERIC: Data

Accumulation, 1969, NAS; Bakus to Brian J. Rothschild, Mar. 16, 1970, Folder Correspondence: Reading File—1970, NAS.

53. William Randolph Taylor to Bakus, Feb. 11, 1970, Folder General 1, 1969–70, NAS.

54. Bakus to Mayr, Dec. 10, 1969, Folder Correspondence: Reading File, 1969, NAS; Bakus to Mayr, Mar. 30, 1970, Folder Correspondence: Reading File—1970, NAS.

55. Voss, "Biological Results of the University of Miami Deep-Sea Expeditions."

56. Bakus to Longhurst, Dec. 15, 1969, Folder Correspondence: Reading File, 1969, NAS.

57. Mayr to Sheffey, Aug. 19, 1969, Folder Biology & Agriculture: CERIC: Data Accumulation, 1969, NAS.

58. Bakus to Mayr, Feb. 26, 1970, Folder Correspondence: Reading File—1970, NAS.

59. Rubinoff, "Sea-Level Canal Controversy," 35.

60. Bakus, "AAAS Sea-Level Canal Symposium, 30 Dec 1969," Jan. 5, 1970, Folder Biology & Agriculture: Interoceanic Canal, Com. on Ecological Research on: Meeting January 15–19, 1970, NAS.

61. Vermeij, *Biogeography and Adaptation*, 259, 266.

62. Kingsland, *Evolution of American Ecology*; Kristin Johnson, "Natural History as Stamp Collecting."

63. Chesher, "Destruction of Pacific Corals"; Sapp, *What Is Natural.*

64. Bakus, "AAAS Sea-Level Canal Symposium, 30 Dec 1969," Jan. 5, 1970, Folder Biology & Agriculture: Interoceanic Canal, Com. on Ecological Research on: Meeting January 15–19, 1970, NAS; David L. Pawson to Bakus, Feb. 5, 1970, Folder General 1, 1969–70, NAS.

65. "Sea-Level Mysteries: Ecology and the Canal," *Science News*, Apr. 11, 1970, 364–65.

66. Rubinoff to Carl J. George, Mar. 24, 1969, RU 108, Box 42, Folder Sea Level Canal Correspondence (1), SIA; Rubinoff and Kropach, "Differential Reactions of Atlantic and Pacific Predators."

67. Graham, Rubinoff, and Hecht, "Temperature Physiology of the Sea Snake."

68. Sheffey to Wallen, Apr. 9, 1969, RU 108, Box 42, Folder Sea Level Canal Correspondence (1), SIA.

69. Rubinoff to Peter S. Hunt, Jan. 20, 1969, RU 108, Box 42, Folder Sea Level Canal Correspondence (1), SIA.

70. CERIC, *Marine Ecological Research*, 2, 188.

71. Victor Cohn, "A-Canal Dealt Blow," *Washington Post*, Apr. 13, 1970.

72. Mayr to Bakus, Apr. 16, 1970, Folder Biology & Agriculture: CERIC 1970, NAS; Stevens to John S. Coleman, Apr. 23, 1970, Folder Biology & Agriculture: CERIC 1970, NAS.

73. Daniel J. Flood to John C. Briggs, Oct. 13, 1969, Folder Studies re Sea-Level Canal, 1969–1972 (2), Box 46, Papers of William Merrill Whitman, DDEL.

74. Daniel J. Flood, "Preconceived Plan for Sea-Level Canal Destroyed: Time for Action on Terminal Lake–Third Locks Plan Has Come," *Congressional Record—House*, May 20, 1970, H4619–H4622; Spear, *Daniel J. Flood*, 75.

75. R. H. Groves to John S. Coleman, May 11, 1970, Folder Biology & Agriculture: CERIC 1970, NAS; Sheffey to Bakus, May 18, 1970, Folder Biology & Agriculture: CERIC 1970, NAS.

76. William A. Newman to John S. Coleman, May 28, 1970, Folder Biology & Agriculture: CERIC 1970, NAS; Howard L. Sanders to Sheffey, June 1, 1970, Folder Biology & Agriculture: CERIC 1970, NAS.

77. Newman, "National Academy of Science Committee," 256.

78. Bayer, Voss, and Robins, *Bioenvironmental and Radiological Safety Feasibility Studies.*

79. U.S. Congress, *Sea-Level Canal Studies*, 374.

80. Voss, "Biological Results of the University of Miami Deep-Sea Expeditions," 54.

81. "Battelle Memorial Institute Report on Possible Effects of a Sea-Level Canal on the Marine Ecology of the American Isthmian Region," quoted in APICSC, *Interoceanic Canal Studies 1970*, 61.

82. "28th Meeting (9–10 July 1970) Transcript," pp. 202–4, Entry A1 36040-C, Container 5, RG 220, NARA; CERIC, *Marine Ecological Research*, 155.

83. "28th Meeting (9–10 July 1970) Transcript," pp. 202–4.

84. "28th Meeting (9–10 July 1970) Transcript," pp. 223–24.

85. "28th Meeting (9–10 July 1970) Transcript," p. 226.

86. Richard Halloran, "Route Is Chosen for New Panama Canal," *New York Times*, Nov. 14, 1970; Thomas O'Toole, "Study Seen Urging 2d Panama Canal," *Washington Post*, Nov. 14, 1970.

87. APICSC, *Interoceanic Canal Studies 1970.*

88. "Transcript of the Executive Session of the 26th Meeting of the APICSC, 12 March 1970," p. 31, Entry A1 36040-C, Container 5, RG 220, NARA.

89. APICSC, *Interoceanic Canal Studies 1970*; U.S. Congress, *Sea-Level Canal Studies*, 302–20.

90. APICSC to the President, Dec. 1, 1970 (8-31-70 version), Folder Final Report of Commission, Entry A1 36040-D, Container 7, RG 220, NARA.

91. Robert M. Sayre to Anderson, Dec. 16, 1970, Box 264, Folder P July–Dec. 1970 (1), Papers of Robert B. Anderson, DDEL. See also Sayre, *Interview with Robert M. Sayre*, by Thomas J. Dunnigan, Oct. 31, 1995.

92. Richard Halloran, "Route Is Chosen for New Panama Canal," *New York Times*, Nov. 14, 1970.

93. Commoner, *Closing Circle*, 60.

94. APICSC, *Interoceanic Canal Studies 1970*, 62.

95. Laycock, *Diligent Destroyers*; Elizabeth Drew, "Dam Outrage—The Story of the Army Engineers," *Atlantic Monthly* (Apr. 1970): 51–62; Morgan, *Dams and Other Disasters.*

96. Boffey, "Sea-Level Canal," 355.

97. Boffey, 358.

98. The full passage is: "The Commission 'leaked' a report to the News media in

which the Academy's report is mentioned, but most of the attention is paid to the report of the Batelle [sic] report about the value of which we entirely agree with you. It is at this point that I got in touch with Mr. Boffey and this, in turn, resulted in his write-up in Science." Mayr to Longhurst, Aug. 11, 1971, Box 20, Folder 1124, Papers of Ernst Mayr, HUA.

99. Mayr to Rubinoff, Nov. 24, 1970, Box 19, Folder 1088, Papers of Ernst Mayr, HUA.

100. Longhurst to Mayr, July 27, 1971, Box 20, Folder 1124, Papers of Ernst Mayr, HUA; Mayr to Longhurst, Aug. 11, 1971.

101. Newman, "National Academy of Science Committee"; Dunson, "Sea Snakes and the Sea Level Canal Controversy"; Rubinoff, "Sea-Level Canal in Panama"; Beeton, *Report of the Committee on Ecological Effects*, 194; Sapp, *Coexistence*, 103.

102. APICSC, *Interoceanic Canal Studies 1970*, V-303.

103. Sapp, *What Is Natural*, 125; Sapp, *Coexistence*, 103.

104. Mayr, *Growth of Biological Thought*, 1.

105. Flood, "1971 Statement before the Subcommittee on Inter-American Affairs," 72; Francis B. Kent, "The Biological Unknowns of a New Panama Canal," *Washington Post*, Jan. 18, 1972. See also Strom Thurmond Statement, Department of State Memorandum of Conversation, July 30, 1971, Congressional Visit—Panama Treaty Negotiations, Box 270, Folder Congressional Meetings, Visits or Telephone Conversations (2), Papers of Robert B. Anderson, DDEL; Associated Press, "Sea Life Presents Problem to Canal," *Sarasota Herald-Tribune*, Dec. 10, 1971; Meredith L. Jones and R. B. Manning, "A Two-Ocean Bouillabaisse Can Result If and When Sea-Level Canal is Dug," *Smithsonian* 2, no. 9 (1971): 12–21.

106. Acting Secretary of State, Memorandum for the President, Dec. 31, 1971, p. 4, Box 269, Folder APICSC 1971–73, Papers of Robert B. Anderson, DDEL.

107. John N. Irwin II, "NSC Under Secretaries Committee Memorandum for the President," June 10, 1971, in *Declassified Documents Reference System*.

108. See also Stine, *Mixing the Waters*; Noll and Tegeder, *Ditch of Dreams*.

Chapter 7. Optioning the Sea-Level Canal for the Energy Crisis

1. Informally named after Panamanian general Omar Torrijos and U.S. president Jimmy Carter, the two pacts are officially titled "The Treaty Concerning the Permanent Neutrality and Operation of the Panama Canal," which grants the United States permanent rights to defend the canal, and "The Panama Canal Treaty," which grants Panama full sovereignty over the waterway; the two accords are commonly called the Panama Canal Treaties. Major studies of the negotiations and ratification process include LaFeber, *Panama Canal*; Jorden, *Panama Odyssey*; Moffett, *Limits of Victory*; Major, *Prize Possession*. On the long-term domestic political and economic effects of the treaties, see Moffett, *Limits of Victory*, chap. 5; Clymer, *Drawing the Line*; Zaretsky, "Restraint or Retreat?"; Maurer and Yu, *Big Ditch*.

2. Glad, *Outsider in the White House*, 93.

3. Moffett, *Limits of Victory*, 40; Linowitz, *Making of a Public Man*, 168.

4. Lindsay-Poland, *Emperors in the Jungle*, 74; Kaufman, *Project Plowshare*, 192.

5. Stine, "Environmental Policy during the Carter Presidency"; Reisner, *Cadillac Desert*, chap. 9; Daynes and Sussman, *White House Politics and the Environment*, chap. 4; Eastman, "Hit List"; Eizenstat, *President Carter*, chap. 12.

6. Stine, "Environmental Policy during the Carter Presidency."

7. Charles Jones, *Trusteeship Presidency.*

8. Kaufman and Kaufman, *Presidency of James Earl Carter*, 32; Godbold, *Jimmy and Rosalynn Carter*, 176; Kaufman, *Plans Unraveled*, 16.

9. See, e.g., Morris, *Jimmy Carter*; Greenberg, *"What the Heck Are You Up To."*

10. An image of the newspaper headline is available at Margaret Kriz Hobson, "Big Finds, Bitter Clashes and NEPA: The Tale of Trans-Alaska," *E&E News*, Aug. 2, 2017, https://www.eenews.net/stories/1060058240.

11. J. G. Phillips, "Alaskan Oil Boom," *Editorial Research Reports* 2 (1969).

12. Coen, *Breaking Ice for Arctic Oil.*

13. Scott Kirsch and Mitchell, "Earth-Moving as the 'Measure of Man,' " 128. See also Scott Kirsch, *Proving Grounds*, 202; "Harbor on North Slope Dug by Nuclear Blast Is Mulled, Teller Says," *Wall Street Journal*, Sept. 25, 1969.

14. Cicchetti, *Alaskan Oil*; Gravel, *Panama Canal*, 40.

15. Peter Coates, *Trans-Alaska Pipeline Controversy*, 178.

16. Turner, *David Brower*, 179.

17. McKloskey, *In the Thick of It*, 104.

18. Peter Coates, *Trans-Alaska Pipeline Controversy*, 189.

19. Coates, 196–206.

20. "Mondale, Walter F.—RNC Quotebooks (1)," Ron Nessen Papers, Gerald R. Ford Presidential Library, https://www.fordlibrarymuseum.gov/library/document/0204/1512122.pdf. See also Peter Coates, *Trans-Alaska Pipeline Controversy*, 241; J. P. Smith, "Alaskan Oil to Begin Flowing Today; Pipeline Oil to Create Glut on West Coast," *Washington Post*, June 20, 1977, A1; "Senate Rejected Mondale's '73 Prediction of Disaster," *Boston Globe*, Mar. 29, 1989.

21. U.S. Department of the Interior, *Final Environmental Impact Statement*, 1:1.

22. Peter Coates, *Trans-Alaska Pipeline Controversy*, 227–47.

23. Liroff, "NEPA Litigation in the 1970s," 316; Dowie, *Losing Ground.*

24. Peter Coates, *Trans-Alaska Pipeline Controversy*, 265; Michael Storper, Laura Baker, and Mary Lou Seaver, "Alaskan Oil: Too Much, Too Soon (Too Bad)," *Not Man Apart* (Apr. 1977), repr. in *Congressional Record*, Feb. 22, 1977, 4876–78.

25. Wallace Turner, "Was This Pipeline Necessary," *New York Times*, June 26, 1977.

26. J. P. Smith, "Alaskan Oil to Begin Flowing Today," *Washington Post*, June 20, 1977.

27. John Jacobs, "Calif. Hit for Stalling Pipeline," *Washington Post*, Aug. 11, 1977.

28. U.S. Congress, *Sea-Level Canal Studies*, 278; Storper et al., "Alaskan Oil," 4877.

29. Lou Cannon, "California Cool to Alaska Oil Terminal," *Washington Post*, July

18, 1977; Wallace Turner, "Was This Pipeline Necessary," *New York Times*, June 26, 1977; Bill Richards, "Energy vs. Environment: Oil, Environment Clash on West Coast," *Washington Post*, Jan. 10, 1977.

30. "Summary of Facts and Opinions Presented to the Atlantic-Pacific Interoceanic Canal Study Commission in Reference to the Future Attractiveness of a Sea-Level Isthmian Canal to the Shippers of Petroleum," Box 65, Folder Oil Company and Shipping Company File, Entry A1 36040-B APICSC Working Files, Studies and Reports, Container 9, RG 220, NARA. See also Associated Press, "Panama Canal Has Lost Strategic Value," *News & Sun-Sentinel* (Fort Lauderdale), Mar. 13, 1988.

31. Ryan, *Panama Canal Controversy*, 83; Department of State, *Final EIS*, 34.

32. Zbigniew Brzezinski to Jimmy Carter, Memorandum: Panama Canal Treaty—Last Decisions, July 28, 1977, Office of the Chief of Staff Files, Hamilton Jordan's Confidential Files, Panama Canal Treaty 6–7/77, Container 36, JCPL.

33. Ryan, *Panama Canal Controversy*, 83; Gravel, *Panama Canal*, 25.

34. Jorden, *Panama Odyssey*, 5.

35. Linowitz, *Making of a Public Man*, 168.

36. *FRUS, 1969–1976*, 22:95.

37. Jorden, *Panama Odyssey*, 289; Major, *Prize Possession*, 348.

38. LaFeber, *Panama Canal*, 190–91; Clymer, *Drawing the Line*.

39. Linowitz, *Making of a Public Man*; Major, *Prize Possession*, 345–49.

40. Daniel J. Flood to Jimmy Carter, Jan. 27, 1977, Folder FO 3-1/Panama Canal 1/20/77–4/13/77, Box FO-15, WHCF Subject File Federal Government—Organization, JCPL.

41. Frisch and Kelly, *Jimmy Carter and the Water Wars*, 39–40; Stine, "Environmental Policy during the Carter Presidency," 184.

42. Flood, "1971 Statement before the Subcommittee on Inter-American Affairs," 72; Francis B. Kent, "The Biological Unknowns of a New Panama Canal," *Washington Post*, Jan. 18, 1972; Spear, *Daniel J. Flood*, 75.

43. Testimony of Strom Thurmond on S. 2330, "A Bill to Provide for the Increase of Capacity and the Improvement of Operations of the Panama Canal," *Congressional Record*, Aug. 2, 1973, S15407.

44. Dunson, "Sea Snakes and the Sea Level Canal Controversy," 518.

45. "Hearings on the Value of the Panama Canal," *Congressional Record*, July 10, 1973, 23091.

46. Turner, *David Brower*; Brower, *For Earth's Sake.*

47. U.S. Congress, *National Outdoor Recreation Programs and Policies*, 298; Alderson, "Interview with George Alderson," by Ted Hudson, 1989; Suisman, "American Environmental Movement's Lost Victory"; Conway, *High-Speed Dreams*, 140, 145; Alderson, *How You Can Influence Congress.*

48. Gary Soucie to Katherine Fletcher, May 20, 1977, Folder FO 3-1/Panama Canal, 1/20/77–9/6/77, Box FO-22, WHCF, Foreign Affairs, JCPL.

49. "Proposed Sea Level Panama Canal Gets Little Support from Shipping Interests," *New York Times*, July 22, 1973.

50. George Alderson to Leonor K. Sullivan, Aug. 14, 1973, Papers of Walter F. Mondale, Minnesota Historical Society; Alderson to Robert L. Leggett, Aug. 14, 1973, Papers of Walter F. Mondale, Minnesota Historical Society.

51. Briggs, "International Symposium," 61.

52. Ortman testimony, U.S. Congress, *Sea-Level Canal Studies*, 369.

53. Dee Arntz, "Fletcher, Kathy (b. 1950)," Essay 9369, HistoryLink.org, Mar. 28, 2010, http://www.historylink.org/File/9369.

54. Walter Pincus, "When a Campaign Vow Crashes into a Pork Barrel," *Washington Post*, Apr. 1, 1977.

55. George Alderson to Katherine Fletcher, Mar. 21, 1977, Folder FO 3-1/Panama Canal, 1/20/77–9/6/77, Box FO-22, WHCF, Foreign Affairs, JCPL.

56. Aron and Smith, "Ship Canals and Aquatic Ecosystems."

57. Fletcher to Alderson, Apr. 6, 1977, Folder FO 3-1/Panama Canal, 1/20/77–9/6/77, Box FO-22, WHCF, Foreign Affairs, JCPL.

58. U.S. Congress, *Sea-Level Canal Studies*, 36.

59. Gravel to Carter, May 5, 1977, Folder FO 3-1/Panama Canal 4/14/77–7/31/77, Box FO-15, WHCF Subject File Federal Government—Organization, JCPL. See also *FRUS, 1977–1980*, 29:64.

60. Gravel, *Panama Canal*, 43.

61. Gravel, 33–43.

62. Alderson to Fletcher, May 16, 1977, Folder FO 3-1/Panama Canal, 1/20/77–9/6/77, Box FO-22, WHCF, Foreign Affairs, JCPL.

63. Gary Soucie to Fletcher, May 20, 1977, Folder FO 3-1/Panama Canal, 1/20/77–9/6/77, Box FO-22, WHCF, Foreign Affairs, JCPL.

64. Fletcher to Alderson, June 23, 1977, Folder 2, Box 21, Series Domestic Policy Staff: Kathy Fletcher's Subject Files, JCPL; Fletcher to Soucie, June 23, 1977, Folder 2, Box 21, Series Domestic Policy Staff: Kathy Fletcher's Subject Files, JCPL. Alderson later transmitted his files to his FOE successor David Ortman, who shared the letter with Congress during the 1978 hearings addressed in chapter 8. David E. Ortman to George D. Moffett, Sept. 10, 1981, Folder 6, Box 7, George D. Moffett Papers, JCPL; Ortman testimony, U.S. Congress, *Sea-Level Canal Studies*, 369. See also "Enemies of the Sea-Level Canal," *New Scientist*, Nov. 3, 1977, 271.

65. Memorandum re: Approved Presidential Activity from Tim Kraft to Frank Moore, 6/30/77: Meeting with Senator Mike Gravel, Folder FO 3-1/Panama Canal 4/14/77–7/31/77, Box FO-15, WHCF Subject File Federal Government—Organization, JCPL; *FRUS, 1977–1980*, 29:64.

66. Briefing Paper prepared by Frank Moore, July 12, 1977, for Meeting with Senator Mike Gravel, July 13, 1977, Folder 7/13/77 [2], Box 31, Office of Staff Secretary Handwriting File, Presidential Files, JCPL.

67. Jimmy Carter, "Remarks on a Question-and-Answer Session at a Public Meeting," July 21, 1977, in *PPPUS*, 1977, 2:1326.

68. Austin Scott, "Carter Visits Floating Oil Rig, Muses on a Sea-Level Canal," *Washington Post*, July 22, 1977; James Nelson Goodsell, "Carter Remarks on Panama

Stir Up a Tempest," *Christian Science Monitor*, July 25, 1977; "Question-and-Answer Session with Reporters," July 22, 1977, in *PPPUS, Jimmy Carter*, 1977, 2:1338–39.

69. The $5.29 billion figure is from Gravel's *Panama Canal* (p. 35), which identified the source as a May 5, 1977, letter by Army Corps of Engineers general Ernest Graves. The Panama Canal Company also used the figure of $5.2 billion in testimony before the Committee on Environment and Public Works on July 22, 1977.

70. Graham Hovey, "Carter Talk May Offend Panama," *New York Times*, July 22, 1977.

71. "Question-and-Answer Session with Reporters," July 22, 1977, in *PPPUS, Jimmy Carter*, 1977, 2:1339; Office of the White House Press Secretary, "Remarks of the President and Question and Answer Session at the Airport Hilton Hotel," July 22, 1977, Records of the Domestic Policy Staff, Folder Sea Level Canal, 1971–5/20/80, Box 70, JCPL.

72. Phil Smith to Raph [sic] Kasper, July 27, 1977, Folder Memos and Correspondence–Presidentials, 5/25/77–8/3/77 (2), Box 1, Science & Tech. Advisor to the President—Press, JCPL.

73. Frank Press to Jimmy Carter, Aug. 1, 1977, Folder Memos and Correspondence–Presidentials, 5/25/77–8/3/77 (2), Box 1, Science & Tech. Advisor to the President—Press, JCPL.

74. Jorden, *Panama Odyssey*, 6–7, 16–19, 429–31, 433–37, 453–56; Linowitz, *Making of a Public Man*, 171.

75. *FRUS, 1977–1980*, 29:64.

76. *FRUS, 1977–1980*, 29:64.

77. Memorandum for Hamilton Jordan from the Vice President, July 29, 1977, 2, Papers of Walter F. Mondale, Minnesota Historical Society, available online as page 22 of the PDF at http://www2.mnhs.org/library/findaids/00697/pdfa/00697-00081-1.pdf.

78. Quoted in Eleanor Randolph, "Talk of Another Canal," *Chicago Tribune*, Aug. 14, 1977.

79. Ernest F. Hollings, "The Panama Canal," *The Fritz Hollings Report* (Sept. 1977), p. 3, Papers of Ernest F. "Fritz" Hollings, South Carolina Political Collections, University of South Carolina Libraries, http://digital.tcl.sc.edu/cdm/ref/collection/how/id/245.

80. *FRUS, 1977–1980*, 29:76.

81. Graham Hovey, "Panamanians Say Carter's Idea for Sea-Level Canal Imperiled Talks," *New York Times*, Aug. 25, 1977.

82. *FRUS, 1977–1980*, 29:76.

83. *FRUS, 1977–1980*, 29:76; see also *FRUS, 1977–1980*, 29:76, Tab A, 29:84.

84. Ernest F. Hollings, "The Panama Canal," *The Fritz Hollings Report* (Sept. 1977), p. 3, Papers of Ernest F. "Fritz" Hollings, South Carolina Political Collections, University of South Carolina Libraries, http://digital.tcl.sc.edu/cdm/ref/collection/how/id/245.

85. Larry Pressler, letter to the editor, *Chicago Tribune*, Oct. 12, 1977.

86. "Who Slipped This In?," *Chicago Tribune*, Sept. 10, 1977.

87. "Remarks of the President on Panama Canal Treaties," Feb. 1, 1978, Folder 2, Box 60, National Security Affairs, Brzezinski Material, Country File, Panama, JCPL.

88. Linowitz, *Making of a Public Man*, 170.

89. Graham Hovey, "Panamanians Say Carter's Idea for Sea-Level Canal Imperiled Talks," *New York Times*, Aug. 25, 1977.

90. "Text of a Telegram Sent to President Carter by 11 National Environmental Organizations on September 6, 1977," repr. in U.S. Congress, *Sea-Level Canal Studies*, 370–71.

91. Edward Flattau, "Sea-Level Canal: A Passage to Ecological Disaster," *Chicago Tribune*, Sept. 10, 1977.

92. Jimmy Carter, "The Environment Message to the Congress," May 23, 1977, in *PPPUS*, 1977, 1:967. See also Luther Carter, "Carter Places Environment High on Agenda."

93. Jimmy Carter, "Remarks at the Signing Ceremony at the Pan American Union Building, September 7, 1977," in *PPPUS*, 1977, 2:1543; Graham Hovey, "Carter, Torrijos Sign Canal Pacts in the Presence of Latin Leaders," *New York Times*, Sept. 8, 1977.

94. Graham Hovey, "Panamanians Say Carter's Idea for Sea-Level Canal Imperiled Talks," *New York Times*, Aug. 25, 1977.

95. Carter, "The Environment Message to the Congress," May 23, 1977, in *PPPUS*, 1977, 1:967.

96. Bourne, *Jimmy Carter*, 72–77; Godbold, *Jimmy and Rosalynn Carter*, 66–68.

97. Godbold, *Jimmy and Rosalynn Carter*, 83, 134.

98. Jon Hardheimer, "Yes, That Was a Georgia Governor Speaking," *New York Times*, Jan. 17, 1971, quoted in Godbold, *Jimmy and Rosalynn Carter*, 174.

99. Richard D. Lyons, "House, by 232–131, Kills Carter Plan for Energy Board," *New York Times*, June 28, 1980; Robert D. Hershey Jr., "Blessing or Boondoggle? The $88 Billion Quest for Synthetic Fuels," *New York Times*, Sept. 1, 1980.

100. Stine, "Environmental Policy during the Carter Presidency," 187, 191, 195.

101. See, e.g., Doel, "Constituting the Postwar Earth Sciences"; Kroll, *America's Ocean Wilderness*; Hamblin, *Poison in the Well*; Finley, *All the Fish in the Sea*; Dorsey, *Whales and Nations*.

102. Carter, *An Outdoor Journal*, 6.

Chapter 8. Containing the Panama Canal Treaty's Environmental Fallout

1. Moffett, *Limits of Victory*; Hogan, "Public Opinion and American Foreign Policy."

2. In addition, several of the implementation agreements for specific treaty articles pertained to environmental issues. These included canal water conservation; the continuation of the research facilities of the Smithsonian Tropical Research Institute, U.S. Army Tropic Test Center, and Gorgas Memorial Institute of Tropical and Preventive Medicine; and the removal of hazards from defense sites. Department of State, *Final EIS*; U.S. Department of State, "Panama Canal Treaty, 313–14. On the U.S. military's refusal to meet its treaty obligations regarding the cleanup of munitions, see Lindsay-Poland, *Emperors in the Jungle*, chap. 5.

3. Robinson, "Environmentalist Looks at the Panama Canal Treaties"; Robinson, "Introduction."

4. See, e.g., Peter Coates, *Trans-Alaska Pipeline Controversy*; Ramey, "Calvert Cliffs Campaign"; Dreyfus and Ingram, "National Environmental Policy Act"; Taylor, *Making Bureaucracies Think*; Caldwell, *National Environmental Policy Act*; Clark and Canter, *Environmental Policy and NEPA*.

5. U.S. Council on Environmental Quality, *Fifth Annual Report*, 392, 399–400. See also Stein, "United States Council on Environmental Quality Memorandum"; Weinstein-Bacall, "The Darien Gap Case"; Lindstrom and Smith, *National Environmental Policy Act*, 95–96; Macekura, *Of Limits and Growth*, chap. 5.

6. Department of State, *Final EIS*, 48. See also Lewis H. Diuguid, "U.S. Study of Canal Pacts Cites Peril to Watershed," *Washington Post*, Oct. 10, 1977; Leonard C. Meeker and Don G. Scroggin to William Mansfield III, Sept. 28, 1977, in Department of State, *Final EIS*, P-25.

7. Harold B. Green Jr., Patricia T. Fulton, Charlotte Kennedy, Jeri Steele, and Melvin Borenam to William Mansfield III, Sept. 26, 1977, p. 5, in Department of State, *Final EIS*, P-41.

8. Although Carter nominated her as one of the three members of CEQ on May 16, 1977, she withdrew her name in January 1978 after senators delayed action on her confirmation due to her lobbying activities. "Council on Environmental Quality: Nomination of Marion Edey to Be a Member," May 16, 1977, in *PPPUS, Jimmy Carter, 1977*, 1:885; "Notes on People," *New York Times*, Jan. 28, 1978.

9. Marion Edey to Stuart Eizenstat, Aug. 1, 1977, Folder 6, Box 7, George D. Moffett Papers, JCPL.

10. Frank Press to Jimmy Carter, Aug. 1, 1977, Folder Memos and Correspondence–Presidentials, 5/25/77–8/3/77 (2), Box 1, Science & Tech. Advisor to the President—Press, JCPL.

11. Ortman testimony in U.S. Congress, *Sea-Level Canal Studies*, 369; Beeton, *Report of the Committee on Ecological Effects*.

12. Marion Edey to Frank Press, Aug. 23, 1977, Folder 6, Box 7, George D. Moffett Papers, JCPL. This letter was also reprinted in U.S. Congress, *Sea-Level Canal Studies*, 346–48.

13. Panama Audubon Society Attachment, in Department of State, *Final EIS*, P-33; Bill L. Long to Ambler Moss, Nov. 1, 1977, Folder Environmentalists [and Canal] 7/75–12/77, Box 19, Assistant to the President Joseph Aragon, JCPL.

14. Lewis H. Diuguid, "U.S. Study of Canal Pacts Cites Peril to Watershed," *Washington Post*, Oct. 10, 1977, A30; Department of State, *Final EIS*, Q-4.

15. Department of State, *Final EIS*, P-17, P-24 to P-28.

16. Department of State, P-33, P-39.

17. Springer, "Resolution on the Panama Canal," 336; Meredith L. Jones, "Resolution Concerning the Maintenance of the Existing Fresh Water Barrier in the Panama Canal" (1973), RU 526, Box 13, Folder Panama Canal Alternatives Study, 1973–1983, SIA.

18. The historic water shortage also provided an opportunity for foresters to emphasize links between rural land use, deforestation, and the canal system. See Wadsworth,

"Deforestation: Death to the Panama Canal," 22–25; Carse, "Nature as Infrastructure"; Carse, "Infrastructural Event."

19. See, e.g., Department of State, *Final EIS*, P-24 to P-27.

20. R. Michael Wright to William Mansfield III, Sept. 7, 1977, in Department of State, *Final EIS*, P-37.

21. Department of State, *Final EIS*, P-4.

22. Department of State, P-41.

23. "Unavoidable Adverse Environmental Impacts of Proposed Treaties," *Congressional Record*, Sept. 28, 1977, S15797, in U.S. Congress, *The Proposed Panama Canal Treaties*, 195–201.

24. Department of State, *Final EIS*, 48–49.

25. "Endangered Species in the Canal Zone," in Department of State, *Final EIS*, Tab E.

26. Department of State, *Final EIS*, 21, 39. On RENARE's institutional weakness, see Carse, "Nature as Infrastructure."

27. Department of State, *Final EIS*, 7.

28. Department of State, 32, 33.

29. Beeton, *Report of the Committee on Ecological Effects*.

30. Rubinoff, "Sea-Level Canal in Panama," 261.

31. Beeton, *Report of the Committee on Ecological Effects*, appendix.

32. Beeton, *Report of the Committee on Ecological Effects*, 3 (quote), passim.

33. Press to Carter, Oct. 3, 1977, Folder Memos and Correspondence–Presidentials, 8/12/77–10/3/77, Box 1, Science & Tech. Advisor to the President—Press, JCPL.

34. *FRUS, 1977–1980*, 29:108.

35. Moffett, *Limits to Victory*, 211 (polling data), chap. 5.

36. See, e.g., Moffett, *Limits to Victory*; LaFeber, *Panama Canal*; Clymer, *Drawing the Line*; Zaretsky, "Restraint or Retreat?"

37. Macekura, *Of Limits and Growth*, chap. 5.

38. "Panama Canal Treaty Ratification Campaign, September 15th Meeting," Office of the Chief of Staff Files, Hamilton Jordan's Confidential Files, Panama Canal Treaty 9/77, Container 36, JCPL.

39. Walter Sullivan, "Sea-Level Canal Could Imperil Marine Life at Either End, Biologists Say," *New York Times*, Oct. 10, 1977. See also Walter Sullivan, "Panama Canal: What if Sea Snakes and Starfish Change Oceans?" *New York Times*, Dec. 13, 1970.

40. "Enemies of the Sea-Level Canal," *New Scientist*, Nov. 3, 1977, 271. See also "Environmentalists Barge In on Panama Canal Treaty," *New Scientist*, Nov. 3, 1977, 270–271.

41. Brower, "We Cannot Stand By Silent," 1.

42. Brower quoted in "Enemies of the Sea-Level Canal," *New Scientist*, Nov. 3, 1977, 271.

43. Robert Goodland, "Triple Threat to Panama's Ecology," *Washington Post*, Dec. 10, 1977; "Goodland to World Bank," *Cary Arboretum Newsletter* 4 (Jan.–Feb. 1978): 4. The *Post* article was based on Goodland's peer-reviewed study "Panamanian Development and the Global Environment," *Oikos* 29 (1977): 195–208. On the Bayano dam,

see Wali Alaka, "In Eastern Panama, Land Is the Key to Survival," *Cultural Survival Quarterly Magazine*, Sept. 1989.

44. Ross Simons to Joseph W. Aragon, Nov. 18, 1977, Folder Environmentalists [and Canal] 7/75–12/77, Box 19, Assistant to the President Joseph Aragon, JCPL.

45. David E. Ortman to George D. Moffett, Sept. 10, 1981, Folder 6, Box 7, George D. Moffett Papers, JCPL.

46. Jim Barnes to David Ortman, Chap Barnes, Tom Stoel, Bill Bulter, Michael Wright, Toby Cooper, Lew Regenstein, Dec. 14, 1977, Folder Environmentalists [and Canal] 7/75–12/77, Box 19, Assistant to the President Joseph Aragon, JCPL.

47. David E. Ortman to George D. Moffett, Sept. 10, 1981, Folder 6, Box 7, George D. Moffett Papers, JCPL.

48. Ortman, "Mingling the Two Oceans," in U.S. Congress, *Sea-Level Canal Studies*, 364.

49. Ortman, 368.

50. Ortman, 369.

51. McKloskey, *In the Thick of It*, chap. 11.

52. The summary findings of the cases, *Sierra Club v. Coleman* 405 F. Supp. 53 (1975) and 421 F. Supp. 63 (1976), are available at http://law.justia.com/cases/federal/district-courts/FSupp/405/53/1432761/ and http://law.justia.com/cases/federal/district-courts/FSupp/421/63/1769598/.

53. Miller, "Minding the Gap"; Ficek, "Imperial Routes."

54. Editor's note, "Statement on the Panama Canal Treaties and Environmental Protection," *Sierra* (Apr. 1978): 24; Bill L. Long to Ambler Moss, Nov. 1, 1977, Folder Environmentalists [and Canal] 7/75–12/77, Box 19, Assistant to the President Joseph Aragon, JCPL; James N. Barnes to Ambler Moss, Dec. 1, 1977, Folder Environmentalists [and Canal] 7/75–12/77, Box 19, Assistant to the President Joseph Aragon, JCPL; Jim Barnes to David Ortman, Chap Barnes, Tom Stoel, Bill Bulter, Michael Wright, Toby Cooper, Lew Regenstein, Dec. 14, 1977, Folder Environmentalists [and Canal] 7/75–12/77, Box 19, Assistant to the President Joseph Aragon, JCPL.

55. *FRUS, 1977–1980*, 29:127.

56. Jim Barnes to David Ortman, Chap Barnes, Tom Stoel, Bill Bulter, Michael Wright, Toby Cooper, Lew Regenstein, Dec. 14, 1977, Folder Environmentalists [and Canal] 7/75–12/77, Box 19, Assistant to the President Joseph Aragon, JCPL.

57. Warren Christopher, "Statement on the Panama Canal Treaties and Environmental Protection," Jan. 1978, repr. in *FRUS, 1977–1980*, 29:127 (emphasis in original).

58. Christopher, "Statement on the Panama Canal Treaties and Environmental Protection."

59. Robinson, "Introduction," 238.

60. Christopher, "Statement on the Panama Canal Treaties and Environmental Protection."

61. Robinson, "Environmentalist Looks at the Panama Canal Treaties."

62. Robinson, "Extraterritorial Environmental Protection Obligations," 270. On the

broader role of environmental NGOs in forcing foreign policy agencies to comply with NEPA during the 1970s, see Macekura, *Of Limits and Growth*, chap. 5.

63. Robinson, "Environmentalist Looks at the Panama Canal Treaties," 25–26.

64. *FRUS, 1977–1980*, 29:133.

65. LaFeber, *Panama Canal*, 178–79.

66. Zaretsky, "Restraint or Retreat?," 561.

67. Representative Gene Snyder, in U.S. Congress, *Sea-Level Canal Studies*, 32.

68. U.S. Congress, *Sea-Level Canal Studies*, 36, 38, 48, 42.

69. Strong, "Jimmy Carter and the Panama Canal Treaties," 272.

70. U.S. Congress, *Sea-Level Canal Studies*, 41.

71. U.S. Congress, 38.

72. Stephen Haycox quoted in David Westphal, "Gravel was a Maverick in the '70s, and His Politickin' Hasn't Changed," *Macon* (Georgia) *Telegraph*, Jan. 2, 2008.

73. U.S. Congress, *Sea-Level Canal Studies*, 244–45.

74. U.S. Congress, 374–75.

75. "A Tropical Science Legacy," Smithsonian Tropical Research Institute, Apr. 27, 2018, https://stri.si.edu/story/tropical-science-legacy.

76. Newman, "The National Academy of Science Committee on the Ecology of the Interoceanic Canal," repr. in U.S. Congress, *Sea-Level Canal Studies*, 350–62; John McCosker to Alfred Beeton, Aug. 29, 1977, repr. in U.S. Congress, *Sea-Level Canal Studies*, 344–45; McCosker and Dawson, "Biotic Passage through the Panama Canal."

77. Voss, "Panama Sea-Level Canal—II."

78. U.S. Congress, *Sea-Level Canal Studies*, 309–10, 315–19; Interoceanic Canal Study Act, H.R. 13176, 95th Congress (1977–78), https://www.congress.gov/bill/95th-congress/house-bill/13176?s=1&r=20.

79. U.S. Congress, *Sea-Level Canal Studies*, 310.

80. U.S. Congress, 316.

81. U.S. Congress, 315, 320.

82. *Wall Street Journal*, Dec. 1, 1970 cited in Flood, "The Monroe Doctrine, Latin America and Panama Canal," *CR House*, Feb. 9, 1971, 2265, https://www.govinfo.gov/content/pkg/GPO-CRECB-1971-pt2/pdf/GPO-CRECB-1971-pt2-6-2.pdf; U.S. Congress, *Sea-Level Canal Studies*, 107.

83. U.S. Congress, *Sea-Level Canal Studies*, 323.

84. Sheffey interview by Jorden, May 8, 1979, p. 22, Box 23, Personal Papers of William J. Jorden, LBJL.

85. U.S. Congress, *Sea-Level Canal Studies*, 319.

86. For related insights see especially Heffernan, "Bringing the Desert to Bloom," 108; Scott Kirsch, *Proving Grounds*, 8; Peyton, *Unbuilt Environments*; Sutter, "The World with Us"; Scoville, "Hydraulic Society and a 'Stupid Little Fish.' "

87. Rubinoff, "Sea-Level Canal in Panama."

88. "A Tropical Science Legacy," Smithsonian Tropical Research Institute, Apr. 27, 2018, https://stri.si.edu/story/tropical-science-legacy.

89. Wilson, *Nature Revealed*, 595.

90. Edward O. Wilson, "The Conservation of Life," *Harvard Magazine*, 1974, 29, 31, repr. in Wilson, *Nature Revealed*, 595–602; Wilson and Willis, "Applied Biogeography."

91. Wilson quoted in Claudia Dreifus, "At 90, E. O. Wilson Still Thrives on Being a Scientific Provocateur," *Quanta Magazine*, May 15, 2019. See also Sapp, *Coexistence*; Raby, *American Tropics*.

92. Leslie A. Pray, "Ernst Mayr Dies," *The Scientist*, Feb. 4, 2005.

93. Zaretsky, "Restraint or Retreat?"

Conclusion. Remembering the Unbuilt Canal

1. Robert B. Anderson, Mar. 17, 1966, meeting, Office Files of Harry McPherson, Box 12, Folder Panama Canal, LBJL.

2. Brower et al. to Jimmy Carter, Jan. 30, 1979, 84–85.

3. Office of Technology Assessment, "Environmental Issues Affecting the Panama Canal."

4. Charlotte Elton, "Japan and Panama: Who Is Setting the Agenda?," 9. See also Charlotte Elton, "Japan and Panama: The Role of the Panama Canal."

5. Henry Scott Stokes, "Japan Is Hoping to Build a New Canal in Panama," *New York Times*, Mar. 26, 1980.

6. Charlotte Elton, "Japan and Panama: Who Is Setting the Agenda?" 8; Henry Scott Stokes, "Japan Is Hoping to Build a New Canal," *New York Times*, Mar. 26, 1980; Geoffrey Murray, "New Panama Canal: Shigeo Nagano Says He Can Do It," *Christian Science Monitor*, July 29, 1980.

7. "The Second Panama Canal Project."

8. Stephen Kinzer, "A Century Later, Idea for a Sea-Level Canal Revived," *Boston Globe*, Feb. 3, 1980; William Chapman, "Japan, U.S. Mull New Panama Canal," *Washington Post*, Mar. 27, 1980.

9. *FRUS, 1977–1980*, 29:266.

10. *FRUS, 1977–1980*, 29:264.

11. Mike Gravel interview by Jorden, Mar. 28, 1979, Box 22, Personal Papers of William J. Jorden, LBJL; Morgan quoted in Stephen Kinzer, "A Century Later, Idea for a Sea-Level Canal Revived," *Boston Globe*, Feb. 3, 1980.

12. Gravel interview by Jorden, LBJL.

13. Warren Christopher interview by Jorden, Box 21, Personal Papers of William J. Jorden, LBJL; Wallace Turner, "Gravel Loses a Bitter Fight in Senate Primary in Alaska," *New York Times*, Aug. 28, 1980.

14. Eric Pace, "Panama Oil Pipeline Job Is Assigned," *New York Times*, Mar. 19, 1981; Suman, "Socioenvironmental Impacts of Panama's Trans-Isthmian Oil Pipeline." Until its closure in 1996, the TPP carried 2.7 billion barrels of Alaska North Slope crude; the pipeline was reopened in 2003 to transport Ecuadorian oil, and

its flow direction reversed in 2008. Sandy Fielden, "The Crude from Transpanama," *RBN Energy Daily Blog*, Aug. 15, 2013, https://rbnenergy.com/the-crude-from-transpanama-pipeline-shipments-from-the-gulf-to-the-pacific-coasts.

15. Luther Carter, "Pipeline Problems Exacerbate West Coast Oil Surplus"; Iver Peterson, "Sohio Cancels a Pipeline to Carry Alaskan Oil from Coast to Texas," *New York Times*, Mar. 14, 1979; G. P. Smith, "Sohio Dropped Pactex Pipeline for Profit Gains, Experts Say," *Washington Post*, Mar. 16, 1979. A similar 1980s-era initiative to revive the project was also unsuccessful.

16. U.S. General Accounting Office, *Establishment of Commission to Study Sea-Level Canal and Alternatives*; Joan Donoghue, "Japan-Panama-United States," Clyde Haberman, "U.S., Japan and Panama Plan Study of Canal," *New York Times*, Sept. 5, 1985.

17. Leschine, "Panamanian Sea-Level Canal."

18. "The Second Panama Canal Project," 307.

19. Ken Wells, "Think Tank Thinks Gigantic Thoughts, and Expensive Ones," *Wall Street Journal*, Aug. 28, 1986; "Col. John Sheffey, 70, Dies; Expert on the Panama Canal," *Washington Post*, Nov. 24, 1989.

20. Elena Lombardo, "Panama Canal Alternatives Study Environmental Assessment and Biological Inventory: Historical Perspective," July 8, 1987, RU 526, Box 14, Folder Panama Canal Alternatives Studies 1987, SIA. See also Edward Flattau, "Ecological Dangers of a Sea-Level Canal," *Chicago Tribune*, Feb. 20, 1990.

21. Ross B. Simons to David Sciacchitano, July 5, 1990, RU 526, Box 14, Folder Panama Alternatives Study 1990, SIA.

22. Hayashi and Prescott, "The 1990s in Japan."

23. Kim Keisling quoted in Thomas M. Defrank, "The Canal Zone Is Paradise Lost," *New York Daily News*, Nov. 19, 1999. See also, e.g., Stephen Kinzer, "The Shift is Painful for Panama Zonians," *Boston Globe*, Feb. 18, 1982; Darryl Fears, "For Some, Panama Canal Treaty Symbolizes a Paradise Lost," *Washington Post*, Dec. 31, 1999; Niko Price, " 'Zonians' Mourn Dying Society," *Tulsa World*, Aug. 9, 1998; Lindsay-Poland, *Emperors in the Jungle*, 174.

24. Lindsay-Poland, *Emperors in the Jungle*, chap. 5; Heckadon-Moreno, "Light and Shadows"; Carse, *Beyond the Big Ditch*.

25. Maurer and Yu, *Big Ditch*.

26. Comision de Estudio de las Alternativas al Canal de Panama, *Informe Final de la Comision para el Estudio de las Alternativas al Canal de Panama* (1993) and *Proceedings of the Universal Congress of the Panama Canal* (Sept. 7–10, 1997), cited in Brooks, "Economic Growth, Ecological Limits, and the Expansion of the Panama Canal," 24. See also Jaén Suárez, *Hombres y Ecología en Panamá*, chap. 5.

27. Marc Lacey, "Panamanians Vote Overwhelmingly to Expand Canal," *New York Times*, Oct. 23, 2006.

28. Gonzalez, "Environmental Impact Assessment in Post-Colonial Societies," 343.

29. Steven Mufson, "An Expanded Panama Canal Opens for Ships," *Chicago Tribune*, June 2, 2016.

30. Andrea Gawrylewski, "Opening Pandora's Locks," *The Scientist*, Oct. 2007, http://www.the-scientist.com/?articles.view/articleNo/25464/title/Opening-Pandora-s-Locks/; Ruiz, Torchin, and Grant, "Using the Panama Canal to Test Predictions"; "Smithsonian Celebrates Panama Canal Expansion!" *Smithsonian Insider*, June 28, 2016, http://insider.si.edu/2016/06/smithsonian-celebrates-panama-canal-expansion/.

31. See, e.g., Schlöder et al., "Pacific Bivalve *Anomia peruviana*"; Ros et al., "The Panama Canal and the Transoceanic Dispersal of Marine Invertebrates.

32. Freestone, Ruiz, and Torchin, "Stronger Biotic Resistance in Tropics."

33. Geburzi and McCarthy, "How Do They Do It?"; Chan and Briski, "Overview of Recent Research in Marine Biological Invasions."

34. Muirhead et al., "Projected Effects of the Panama Canal Expansion."

35. Marco Evers, "Russia Moves to Boost Arctic Shipping," *Spiegel Online*, Aug. 22, 2013.

36. McKeon et al., "Melting Barriers to Faunal Exchange," 465. See also A. Whitman Miller and Ruiz, "Arctic Shipping and Marine Invaders"; Mollie Bloudoff-Indelicato, "If Atlantic and Pacific Sea Worlds Collide, Does That Spell Catastrophe?," *Smithsonian*, Nov. 30, 2015; Cheryl Katz, "Alien Waters: Neighboring Seas Are Flowing into a Warming Arctic Ocean," *Yale Environment 360*, May 10, 2018.

37. "China COSCO Shipping Wins Draw for First Transit through Expanded Panama Canal," *MENA Report*, May 6, 2016; Jenny W. Hsu, "U.S. LNG for China Arrives via Panama Canal," *Wall Street Journal*, Aug. 24, 2016; Ryan Collins and Naureen S. Malik, "A First for Panama Canal: Three LNG Tankers Crossed in a Day," *Bloomberg*, Apr. 18, 2018; Mason Hamilton, "Panama Canal Expansion Allows More Transits of Propane and Other Hydrocarbon Gas Liquids," U.S. Energy Information Administration, Apr. 29, 2019, https://www.eia.gov/todayinenergy/detail.php?id=39272.

38. "Nicaragua Canal Plan Not a Joke," *BBC News*, June 26, 2013.

39. Huete-Pérez, Tundisi, and Alvarez, "Will Nicaragua's Interoceanic Canal Result in an Environmental Catastrophe"; Huete-Pérez, Meyer, and Alvarez, "Rethink the Nicaragua Canal"; Huete-Pérez et al., "Scientists Raise Alarms"; Huete-Pérez et al., "Critical Uncertainties and Gaps"; Härer, Torres-Dowdall, and Meyer, "Imperiled Fish Fauna." For alternative viewpoints, see Condit, "Extracting Environmental Benefits."

40. Suzanne Daley, "Lost in Nicaragua, a Chinese Tycoon's Canal Plan," *New York Times*, Apr. 4, 2016; Stephen Gibbs and Lucinda Elliott, "China Puts Nicaraguan Canal Plan on Hold," *Sunday Times* (London), June 19, 2017; Andréas Oppenheimer, "Four Years Later, Nicaragua's $40 Billion Interoceanic Canal Remains a Pipe Dream," *Miami Herald*, July 5, 2017; "Nicaragua's US$50B Rival to Panama Canal 'Going Ahead Slowly' as Funding Evaporates and Chinese Investor Keeps Low Profile," *South China Morning Post*, Feb. 22, 2018; Nicholas Muller, "Nicaragua's Chinese-Financed Canal Project Still in Limbo," *Diplomat*, Aug. 20, 2019.

41. Fred Pearce, "Mega-Canals Could Slice through Continents for Giant Ships," *New Scientist*, Apr. 11, 2017.

42. William Laurance, "Is the Global Era of Massive Infrastructure Projects Coming

to an End?," *Yale Environment 360*, July 10, 2018; McCall and Taylor, "Nicaragua's 'Grand Canal,'" 195. See also Flyvbjerg, "Survival of the Unfittest."

43. Henry Fountain, "Water Levels Drop at Panama Canal, as Climate Change Alters Weather Patterns," *New York Times*, May 18, 2019; Carse, "Infrastructural Event."

44. Covich, "Projects That Never Happened"; Covich, "Frank Golley's Perspectives."

BIBLIOGRAPHY

Manuscript Collections

Dwight D. Eisenhower Presidential Library, Abilene, Kans.
  Papers of Robert B. Anderson
  Papers of William Merrill Whitman
Harvard University Archives, Cambridge, Mass.
  Papers of Ernst Mayr
Jimmy Carter Presidential Library, Atlanta, Ga.
  Assistant to the President Joseph Aragon
  George D. Moffett Papers
  National Security Affairs, Brzezinski Material, Country File
  Office of the Chief of Staff Files
  Records of the Domestic Policy Staff
  Science and Technology Advisor to the President
  White House Central File
Library of Congress, Washington, D.C.
  Declassified Documents Reference System
Lyndon Baines Johnson Presidential Library, Austin, Tex.
  National Security File, Country File: Latin America—Nicaragua, Panama
  National Security File, Files of Charles E. Johnson
  Office Files of Harry McPherson
  Personal Papers of William J. Jorden
National Academy of Sciences Archives, Washington, D.C.
  Division of Biology and Agriculture Collection, Committee on Ecological Research for Interoceanic Canal, 1969–1970
Smithsonian Institution Archives, Washington, D.C.
  RU 99, Office of the Secretary, Records, 1964–1971
  RU 108, Assistant Secretary for Science, Records, 1963–1973
  RU 136, National Museum of Natural History, Dept. of Vertebrate Zoology, Departmental Records, 1954–1970
  RU 470, Contracts Office, Records, 1953–1990
  RU 526, Assistant Secretary for Research, Records, circa 1973–1990

U.S. National Archives and Records Administration, College Park, Md.
RG 220, Records of the Atlantic-Pacific Interoceanic Canal Study Commission, 1965–1970

Online Archives and Oral Histories

Alderson, George. "Interview with George Alderson." By Ted Hudson, April 15, 1982. In *Sierra Club Oral History Project*. Bancroft Library, University of California, Berkeley, 1989. http://digitalassets.lib.berkeley.edu/roho/ucb/text/sc_nationwide3.pdf.

Graves, Ernest. *Engineer Memoirs: Lieutenant General Ernest Graves, U.S. Army*. Washington, D.C.: U.S. Army Corps of Engineers, 1997. http://www.publications.usace.army.mil/Portals/76/Publications/EngineerPamphlets/EP_870-1-52.pdf.

Irwin, John N., II. *Interview with John N. Irwin II*. By Gordon W. Evans, May 30, 1991. Frontline Diplomacy: The Foreign Affairs Oral History Collection of the Association for Diplomatic Studies and Training. Washington, D.C.: Library of Congress, 1991. https://www.loc.gov/item/mfdipbib000554/.

Papers of Ernest F. "Fritz" Hollings, South Carolina Political Collections, University of South Carolina Libraries, Columbia, S.C. https://digital.library.sc.edu/collections/fritz-hollings-in-his-own-words/.

Papers of Walter F. Mondale, Minnesota Historical Society, St. Paul, Minn. http://www2.mnhs.org/library/findaids/00697.xml.

Sayre, Robert M. *Interview with Robert M. Sayre*. By Thomas J. Dunnigan, Oct. 31, 1995. Frontline Diplomacy: The Foreign Affairs Oral History Collection of the Association for Diplomatic Studies and Training. Washington, D.C.: Library of Congress, 1995. https://www.loc.gov/item/mfdipbib001021/.

Selected Government Documents

Atlantic-Pacific Interoceanic Canal Study Commission. *Interoceanic Canal Studies 1970*. Washington, D.C.: Government Printing Office, 1970.

Beeton, Alfred M. *Report of the Committee on Ecological Effects of a Sea Level Canal, Environmental Studies Board to the Honorable Frank Press*. Washington, D.C.: U.S. National Academy of Sciences, 1977.

"Bibliography of Panama Canal Issues." *Congressional Record—Senate*, July 10, 1967, 18114–19.

Congressional Research Service. *Background Documents Relating to the Panama Canal Prepared for the Committee on Foreign Relations, United States Senate*. Washington, D.C.: Government Printing Office, 1977.

Conn, Stetson, Rose C. Engelman, and Byron Fairchild. *Guarding the United States and Its Outposts*. Washington, D.C.: Government Printing Office, 2000.

Flood, Daniel J. "1971 Statement before the Subcommittee on Inter-American Affairs," Sept. 22, 1971; repr. in *Hearings before the Subcommittee on Separations of Powers of the Committee on the Judiciary, United States Senate, Ninety-Fifth Congress, Part 2, July 29, 1977*, 61–73. Washington, D.C.: Government Printing Office, 1977.

Flood, Daniel J. *Isthmian Canal Policy Questions: Selected Addresses by Representative Daniel J. Flood of Pennsylvania*. Washington, D.C.: Government Printing Office, 1966.

Ford, Harold P. *CIA and the Vietnam Policymakers: Three Episodes, 1962–1968*. N.p.: CIA Center for the Study of Intelligence, 1998.

Gravel, Mike. *The Panama Canal—A Reexamination: A Report to the Committee on Environment and Public Works, United States Senate*. Washington, D.C.: Government Printing Office, 1977. https://catalog.hathitrust.org/Record/002941625.

Graves, Earnest. "Nuclear Excavation of a Sea-level, Isthmian Canal." In *Proceedings of the Third Plowshare Symposium: Engineering with Nuclear Explosives (April 21, 22, 23, 1964)*, 321–34; repr. in U.S. Congress. Joint Committee on Atomic Energy. *Peaceful Applications of Nuclear Explosives*, 365–78. https://catalog.hathitrust.org/Record/000964076.

———, Robert Holmes, Milo Nordyke, Lewis J. Cauthen, and Marvin M. Williamson. *Isthmian Canal Studies—1964*; Appendix 1: Nuclear Excavation Plan. Livermore, Calif.: Lawrence Radiation Laboratory, University of California, Sept. 1964.https://catalog.hathitrust.org/Record/101702267.

Hacker, Barton C. *Fallout from Plowshare: Peaceful Nuclear Explosions and the Environment, 1956–1973*. Lawrence Livermore National Laboratory, Contract LLNL-CONF-464374, 2010.

Irwin, John N., II. "NSC Under Secretaries Committee Memorandum for the President." June 10, 1971. Retrieved from the *Declassified Documents Reference System*. Washington, D.C.: Library of Congress.

Isthmian Canal Studies Board of Consultants. *Report to the Committee on Merchant Marine and Fisheries, House of Representatives, United States Congress, on a Long-Range Program for the Panama Canal [...] June 1, 1960*. Washington, D.C.: Government Printing Office, 1960. https://catalog.hathitrust.org/Record/102005084.

Mehaffey, J. C. *Report of the Governor of the Panama Canal: Isthmian Canal Studies—1947*. Balboa Heights, Canal Zone: n.p., 1947. http://ufdc.ufl.edu/AA00029641/00011.

Morgenstern, Oskar, and Klaus-Peter Heiss. *General Report on the Economics of the Peaceful Uses of Underground Nuclear Explosions*. Princeton: Mathematica, Aug 31, 1967. https://www.osti.gov/servlets/purl/4289629.

Panama Canal Company. *The Panama Canal: The Third Locks Project*. Balboa Heights, Canal Zone: n.p., 1941. http://ufdc.ufl.edu/AA00019286/00001.

Panama Canal Company and Canal Zone Government. *Annual Report: Fiscal Year Ended June 30, 1965*. Washington, D.C.: Government Printing Office, 1965.

Rea, Kennedy F., and Marcellus C. Shield. *Statements for the Seventieth Congress, Second Session: Appropriations, Budget Estimates, Etc.* Washington, D.C.: Government Printing Office, 1929.

Report of the Board of Consulting Engineers for the Panama Canal. Washington, D.C.: Government Printing Office, 1906. https://catalog.hathitrust.org/Record/001515101.

Rubinoff, Ira. Statement. *Hearings before the Subcommittee on the Panama Canal of the Committee on Merchant Marine and Fisheries Committee, House of Representatives, Ninety-Fifth Congress, First Session on C. Z. Biological Area Authorization, March 22, 1977*, 15–22. Washington, D.C.: Government Printing Office, 1977.

Teller, Edward. "The Plowshare Program." In *Proceedings of the Second Plowshare Symposium, Part I: Phenomenology of Underground Nuclear Explosions*, edited by Lawrence Radiation Laboratory-Livermore and AEC-San Francisco Operations Office, 8–13. N.p.: U.S. Atomic Energy Commission, 1959. https://catalog.hathitrust.org/Record/007842421.

Thompson, Theos J. "Improving the Quality of Life—Can Plowshare Help?" In Vol. 1 of *Symposium on Engineering with Nuclear Explosives, Las Vegas, NV, 14–16 Jan 1970: Proceedings*, 1–4. American Nuclear Society and U.S. Atomic Energy Commission, May 1970. https://catalog.hathitrust.org/Record/102756714.

U.S. Congress. House. *National Outdoor Recreation Programs and Policies: Hearings before the Subcommittee on National Parks and Recreation of the Committee on Interior and Insular Affairs, House of Representatives [. . .] March 13, 15, 16, 22, and 23, 1973.* Washington, D.C.: Government Printing Office, 1973.

U.S. Congress. House. *Sea-Level Canal Studies: Hearings before the Subcommittee on the Panama Canal of the Committee on Merchant Marine and Fisheries, House of Representatives, Ninety-Fifth Congress on H.R. 10087 and H.R. 13176, June 21, 27, 28, 1978.* Washington, D.C.: Government Printing Office, 1978. http://ufdc.ufl.edu/AA00006070/00001.

U.S. Congress. Joint Committee on Atomic Energy. *Biological and Environmental Effects of Nuclear War. Hearings before the Special Subcommittee on Radiation of the Joint Committee on Atomic Energy, Congress of the United States, Eighty-Sixth Congress [...] June 22, 23, 25, 25, and 26, 1959.* Washington, D.C.: Government Printing Office, 1959. https://catalog.hathitrust.org/Record/001560661.

U.S. Congress. Joint Committee on Atomic Energy. *Peaceful Applications of Nuclear Explosives—Plowshare. Hearing before the Joint Committee on Atomic Energy, Congress of the United States, Eighty-Ninth Congress [...] January 5, 1965.* Washington, D.C.: Government Printing Office, 1964. https://catalog.hathitrust.org/Record/000964076.

U.S. Congress. Senate. *Authorizing the President to Appoint a Commission to Study the Feasibility of, and Most Suitable Site for, the Second Interoceanic Canal Connecting the Atlantic and Pacific Oceans: Report (to Accompany S. 2701).* Washington, D.C.: Government Printing Office, 1964.

U.S. Congress. Senate. *Nuclear Test-Ban Treaty: Hearings before the Committee on Foreign Relations, United States Senate, Eighty-Eighth Congress, First Session, on [. . .] the Treaty Banning Nuclear Weapon Tests in the Atmosphere, in Outer Space, and Underwater, Signed at Moscow on August 5, 1963 [...] August 12, 13, 14, 15, 19, 20, 21, 22, 23, 26, and 27, 1963*. Washington, D.C.: Government Printing Office, 1963. https://catalog.hathitrust.org/Record/100666977.

U.S. Congress. Senate. *Second Transisthmian Canal: Hearings before the Committee on Commerce, United States Senate, Eighty-Eighth Congress, Second Session, on S. 2428, a Bill to Authorize a Study of Means of Increasing the Capacity and Security of the Panama Canal, and for Other Purposes; and S. 2497, a Bill to Provide for an Investigation and Study to Determine a Site for the Construction of a Sea Level Interoceanic Canal Through the American Isthmus. March 3 and 4, 1964*. Washington, D.C.: Government Printing Office, 1964.

U.S. Congress. Senate. *The Proposed Panama Canal Treaties: A Digest of Information; Prepared for the Committee on the Judiciary, United States Senate, by Its Subcommittee on Separation of Powers*. Washington, D.C.: Government Printing Office, 1978.

U.S. Council on Environmental Quality. *Fifth Annual Report*. Washington, D.C.: Government Printing Office, 1974.

U.S. Department of the Interior and U.S. Federal Task Force on Alaskan Oil Development. *Final Environmental Impact Statement: Proposed Trans-Alaska Pipeline*. Vol. 1. Washington, D.C.: Government Printing Office, 1972.

U.S. Department of State. *Final Environmental Impact Statement for the New Panama Canal Treaties*. Washington, D.C.: Government Printing Office, 1977. https://catalog.hathitrust.org/Record/007474489.

U.S. Department of State. *Foreign Relations of the United States*. history.state.gov.

———. *Foreign Relations of the United States, 1948*. Vol. 9, *The Western Hemisphere*. Edited by Almon R. Wright, Velma Hastings Cassidy, and David H. Stauffer. Washington, D.C.: Government Printing Office, 1972. https://history.state.gov/historicaldocuments/frus1948v09.

———. *Foreign Relations of the United States, 1955–1957*. Vol. 7, *American Republics: Central and South America*. Edited by Edith James, N. Stephen Kane, Robert McMahon, and Delia Pitts. Washington, D.C.: Government Printing Office, 1988. https://history.state.gov/historicaldocuments/frus1955-57v07.

———. *Foreign Relations of the United States, 1961–1963*. Vol. 12, *American Republics*. Edited by Edward C. Keefer, Harriet Dashiell Schwar, and W. Taylor Fain III. Washington, D.C.: Government Printing Office, 1996. https://history.state.gov/historicaldocuments/frus1961-63v12.

———. *Foreign Relations of the United States, 1964–1968*. Vol. 11, *Arms Control and Disarmament*. Edited by Evans Gerakas, David S. Patterson, and Carolyn B. Yee. Washington, D.C.: Government Printing Office, 1997. https://history.state.gov/historicaldocuments/frus1964-68v11.

———. *Foreign Relations of the United States, 1964–1968*. Vol. 31, *South and Central America; Mexico*. Edited by David C. Geyer and David H. Herschler. Washington, D.C.: Government Printing Office, 2004. https://history.state.gov/historicaldocuments/frus1964-68v31.

———. *Foreign Relations of the United States, 1969–1976*. Vol. E–10, *Documents on American Republics, 1969–1972*. Edited by Douglas Kraft and James Siekmeier. Washington, D.C.: Government Printing Office, 2009. https://history.state.gov/historicaldocuments/frus1969-76ve10.

———. *Foreign Relations of the United States, 1969–1976*. Vol. 22, *Panama, 1973–1976*. Edited by Bradley L. Coleman, Alexander O. Poster, and James F. Siekmeier. Washington, D.C.: Government Printing Office, 2015. https://history.state.gov/historicaldocuments/frus1969-76v22.

———. *Foreign Relations of the United States, 1977–1980*. Vol. 29, *Panama*. Edited by Laura R. Kolar. Washington, D.C.: Government Printing Office, 2016. https://history.state.gov/historicaldocuments/frus1977-80v29.

U.S. Department of State. "Panama Canal Treaty: Implementation of Article IV, Use of Defense Sites." In *United States Treaties and Other International Agreements*. Vol. 33, Part I, 1979–1981, 313–14. Washington, D.C.: Government Printing Office, 1987.

U.S. General Accounting Office. *Briefing Report to the Honorable Webb Franklin, House of Representatives: Panama Canal; Establishment of Commission to Study Sea-Level Canal and Alternatives*. Washington, D.C.: U.S. General Accounting Office, 1986. https://catalog.hathitrust.org/Record/011411182.

U.S. Office of the Federal Register. National Archives and Records Service. General Services Administration. *Public Papers of the Presidents of the United States: Jimmy Carter, 1977*. Vols. 1 and 2. Washington, D.C.: Office of the Federal Register, National Archives and Records Service, 1977. https://catalog.hathitrust.org/Record/004732130.

———. *Public Papers of the Presidents of the United States: Lyndon B. Johnson, 1963–64*. Vols. 1 and 2. Washington, D.C.: Office of the Federal Register, National Archives and Records Service, 1965. https://catalog.hathitrust.org/Record/004730949.

———. *Public Papers of the Presidents of the United States: Lyndon B. Johnson, 1965*. Vol. 1. Washington, D.C.: Office of the Federal Register, National Archives and Records Service, 1966. https://catalog.hathitrust.org/Record/004730960.

———. *Public Papers of the Presidents of the United States: Lyndon B. Johnson, 1966*. Vol. 1. Washington, D.C.: Office of the Federal Register, National Archives and Records Service, 1967. https://catalog.hathitrust.org/Record/004731549.

———. *Public Papers of the Presidents of the United States: Lyndon B. Johnson, 1968–1969*. Vol. 2. Washington, D.C.: Office of the Federal Register, National Archives and Records Service, 1970. https://catalog.hathitrust.org/Record/004731573.

———. *Public Papers of the Presidents of the United States: Richard Nixon, 1969*. Washington, D.C.: Office of the Federal Register, National Archives and Records Service, 1970. https://catalog.hathitrust.org/Record/004731731.

———. *Public Papers of the Presidents of the United States: Richard Nixon, 1970*. Washington, D.C.: Office of the Federal Register, National Archives and Records Service, 1971. https://catalog.hathitrust.org/Record/004731750.

———. *Public Papers of the Presidents of the United States: Richard Nixon, 1971.* Washington, D.C.: Office of the Federal Register, National Archives and Records Service, 1972. https://catalog.hathitrust.org/Record/004731800.

U.S. Office of Technology Assessment. "Environmental Issues Affecting the Panama Canal: Working Paper Prepared for House Committee on Merchant Marine and Fisheries, Subcommittee on Panama Canal, Dec. 15, 1978"; repr. in U.S. Congress, *Canal Operation under 1977 Treaty—Part 2: Hearings before the Subcommittee on the Panama Canal of the Committee on Merchant Marine and Fisheries, House of Representatives [. . .]*, 974–1010. Washington, D.C.: Government Printing Office, 1979.

Vortman, L. J. "Excavation of a Sea-Level Ship Canal." In *Proceedings of the Second Plowshare Symposium, May 13–15, 1959, San Francisco, California, Part II: Excavation*, edited by Lawrence Radiation Laboratory-Livermore and AEC-San Francisco Operations Office, 71–88. N.p.: U.S. Atomic Energy Commission, 1959. https://catalog.hathitrust.org/Record/007842422.

Wadsworth, Frank. "Deforestation: Death to the Panama Canal." In *Proceedings of the U.S. Strategy Conference on Tropical Deforestation*. Washington: U.S. Department of State and U.S. Agency for International Development, 1978.

Walker, John G. *Report of the Isthmian Canal Commission, 1899–1901.* Washington, D.C.: Government Printing Office, 1901. https://catalog.hathitrust.org/Record/012142615.

Werth, Glenn C. "Closing Remarks." In Vol. 2 of *Symposium on Engineering with Nuclear Explosives, Las Vegas, NV, 14–16 Jan 1970: Proceedings*, 1771–75. American Nuclear Society and U.S. Atomic Energy Commission, May 1970. https://catalog.hathitrust.org/Record/102756714.

Wolfe, John. "The Ecological Aspects of Project Chariot." In *Proceedings of the Second Plowshare Symposium, May 13–15, 1959, San Francisco, California, Part II: Excavation*, edited by Lawrence Radiation Laboratory-Livermore and AEC-San Francisco Operations Office, 60–66. N.p.: U.S. Atomic Energy Commission, 1959. https://catalog.hathitrust.org/Record/007842422.

Selected Periodicals

Atlantic Monthly
Chicago Tribune
Christian Science Monitor
Los Angeles Times
New Scientist
New York Times
Panama American
Popular Mechanics
The Scientist

Science News
Star & Herald (Panama)
U.S. News & World Report
Wall Street Journal
Washington Post
Washington Star

Published Primary Sources

Alderson, George. *How You Can Influence Congress: The Complete Handbook for the Citizen Lobbyist*. New York: Dutton, 1979.

Allen, Emory Adams. *Our Canal in Panama: The Greatest Achievement in the World's History*. Cincinnati: United States Publishing Company, 1913.

Aron, William I., and Stanford H. Smith. "Ship Canals and Aquatic Ecosystems." *Science* 174 (1971): 13–20.

Baker, Herbert G., and G. Ledyard Stebbins, eds. *The Genetics of Colonizing Species*. New York: Academic Press, 1965.

"Baron von Humboldt's Encouragement, given in 1856, to the United Efforts of all the Maritime Nations for the Construction of a Ship-Passage to the Pacific Ocean [. . .]." In Vol. 1 of *The Writings of William Paterson, Founder of the Bank of England; with Biographical Notices of the Author, His Contemporaries, and His Race*, edited by S. Bannister, 280–83. London: Effingham Wilson, Royal Exchange, 1858.

Battelle Memorial Institute. *Environmental Impact Assessment for Darien Gap Highway from Tocumen, Panama, to Rio Leon, Colombia*. Columbus, Ohio: Battelle Memorial Institute, 1974. https://catalog.hathitrust.org/Record/100981181.

Bayer, Frederick M., Gilbert L. Voss, and C. Richard Robins. *Bioenvironmental and Radiological Safety Feasibility Studies, Atlantic-Pacific Interoceanic Canal: Report on the Marine Fauna and Benthic Shelf-Slope Communities of the Isthmian Region*, No. BMI–171–38. University of Miami, Rosenstiel School of Marine and Atmospheric Sciences, 1970.

Belinfante, A. D., Gustaf Petren, and Navroz Vakil. *Report on the Events in Panama, January 9–12, 1964*. Geneva: International Commission of Jurists, 1964. https://www.icj.org/wp-content/uploads/1964/01/Panama-disturbances-fact-finding-mission-report-1964-eng.pdf.

Bennett, Ira E. *The History of the Panama Canal: Its Construction and Builders*. Washington, D.C.: Historical Publishing Company, 1915.

Bidwell, Charles Toll. *The Isthmus of Panamá*. London: Chapman and Hall, 1865.

Boffey, Philip M. *The Brain Bank of America: An Inquiry into the Politics of Science*. New York: McGraw-Hill, 1975.

———. "Sea-Level Canal: How the Academy's Voice Was Muted." *Science* 171 (1971): 355–58.

Bowman, Waldo O. "Puzzle in Panama." *Engineering News-Record* 138 (May 1, 1947); repr. in *Annual Report of the Board of Regents of Smithsonian Institution, 1947*, 407–28. Washington, D.C.: Government Printing Office, 1948.

Briggs, John C. "An International Symposium: The Sea-Level Panama Canal Controversy." *Defenders of Wildlife News* (Jan. 1973): 60–62.

———. *Marine Zoogeography*. New York: McGraw Hill, 1974.

———. "Panama's Sea-Level Canal." *Science* 162 (1968): 511–13.

———. "Relationship of the Tropical Shelf Regions." In *Proceedings of the International Conference on Tropical Oceanography*, 569–78. University of Miami Institute of Marine Sciences, 1967.

———. "The Sea-Level Panama Canal: Potential Biological Catastrophe." *BioScience* 19 (1969): 44–47.

Brower, David R. "We Cannot Stand By Silent." *Not Man Apart* 7, no. 19 (Nov. 1977): 1.

——— et al. to Jimmy Carter. Jan. 30, 1979; repr. in U.S. Congress, *Panama Canal Implementing Legislation: Hearing and Markup before the Committee on Foreign Affairs, House of Representatives [. . .] April 4 and 5, 1979*, 84–85. Washington, D.C.: Government Printing Office, 1979.

Buechner, Helmut K., and F. Raymond Fosberg. "A Contribution toward a World Program in Tropical Biology." *BioScience* 17 (1967): 532–38.

Buel, C. C. "Piercing the American Isthmus." *Scribner's Monthly* 18 (1879): 268–80.

Carlton, James T. "Transoceanic and Interoceanic Dispersal of Coastal Marine Organisms: The Biology of Ballast Water." *Oceanography and Marine Biology Annual Review* 23 (1985): 313–71.

Carson, Rachel. *Silent Spring*. 1962; repr., Boston: Houghton Mifflin, 2002.

Carter, Jimmy. *An Outdoor Journal: Adventures and Reflections*. New York: Bantam Books, 1988.

Carter, Luther J. "Carter Places Environment High on Agenda." *Science* 196 (1977): 1065.

———. "National Academy of Sciences: Unrest among the Ecologists." *Science* 159 (1968): 287–89.

———. "Pipeline Problems Exacerbate West Coast Oil Surplus." *Science* 201 (1978): 594–98.

———. "Rio Blanco: Stimulating Gas and Conflict in Colorado." *Science* 180 (1973): 844–48.

Chan, Farrah T., and Elizabeta Briski. "An Overview of Recent Research in Marine Biological Invasions." *Marine Biology* 164 (2017): 121.

Chesher, Richard H. "Destruction of Pacific Corals by the Sea Star *Acanthaster planci*." *Science* 165 (1969): 280–83.

———. "Transport of Marine Plankton through the Panama Canal." *Limnology and Oceanography* 13 (1968): 387–88.

Cicchetti, Charles J. *Alaskan Oil: Alternative Routes and Markets*. Baltimore: Johns Hopkins University Press, 1972.

Cole, LaMont C. "Can the World Be Saved?" *BioScience* 18 (1968): 679–84.

Collins, Frederick. "The Isthmus of Darien and the Valley of the Atrato Considered with Reference to the Practicability of an Interoceanic Ship-Canal." *Journal of the American Geographical Society of New York* 5 (1874): 138–65.

Committee on Ecological Research for the Interoceanic Canal. *Marine Ecological Research for the Central American Interoceanic Canal.* Washington, D.C.: National Academy of Sciences, 1970.

Commoner, Barry. *The Closing Circle: Nature, Man, and Technology.* New York: Knopf, 1971.

Condit, Richard. "Extracting Environmental Benefits from a New Canal in Nicaragua: Lessons from Panama." *PLoS Biology* 13 (2015): e1002208.

Cowan, Richard S., D. Davis, P. S. Humphrey, W. H. Klein, P. C. Ritterbush, and S. Shelter. "Meetings." *BioScience* 15 (1965): 607–8.

Dawson, C. E. "Occurrence of an Exotic Eleotrid Fish in Panamá with Discussion of Probable Origin and Mode of Introduction." *Copeia*, no. 1 (1973): 141–44.

de Lesseps, Ferdinand. "The Panama Canal." *Science* 8 (1886): 517–20. Donoghue, Joan E. "Japan-Panama-United States: Exchange of Notes Establishing Commission for the Study of Alternatives to the Panama Canal." *International Legal Materials* 25 (1986): 63–73.

Dreyfus, Daniel A., and Helen M. Ingram. "The National Environmental Policy Act: A View of Intent and Practice." *Natural Resources Journal* 16 (1976): 243–62.

Dunson, William A. "Sea Snakes and the Sea Level Canal Controversy." In *The Biology of Sea Snakes*, edited by William A. Dunson, 517–24. Baltimore: University Park Press, 1975.

Eisenhower, Milton S. *The Wine Is Bitter: The United States and Latin America.* Garden City, N.Y.: Doubleday, 1963.

Elton, Charles S. *The Ecology of Invasions by Animals and Plants.* 1958; repr., Chicago: University of Chicago Press, 2000.

Elton, Charlotte. "Japan and Panama: The Role of the Panama Canal." In *Japan, the United States, and Latin America: Toward a Trilateral Relationship in the Western Hemisphere?*, edited by Barbara Stallings and Gabriel Székely, 210–28. London: Palgrave MacMillan, 1993.

———. "Japan and Panama: Who Is Setting the Agenda?" MIT Japan Program. Paper presented at the XV International Congress of the Latin American Studies Association, Miami, Fla. December 1989. https://dspace.mit.edu/bitstream/handle/1721.1/17081/JP-WP-90-02-22164814.pdf?sequence=1.

Fitz-Roy, Robert. "Considerations on the Great Isthmus of Central America." *Journal of the Royal Geographical Society of London* 20 (1850): 161–89.

Foster, William C. "Risks of Nuclear Proliferation: New Directions in Arms Control and Disarmament." *Foreign Affairs* (1965): 587–601.

Freestone, Amy L., Gregory M. Ruiz, and Mark E. Torchin. "Stronger Biotic Resistance in Tropics Relative to Temperate Zone: Effects of Predation on Marine Invasion Dynamics." *Ecology* 94 (2013): 1370–77.

Geburzi, Jonas C., and Morgan L. McCarthy. "How Do They Do It?—Understanding the Success of Marine Invasive Species." In *YOUMARES 8—Oceans across Boundaries: Learning from Each Other*, edited by Simon Jungblut, Viola Liebich, and Maya Bode, 109–24. Springer, Cham, 2018.

Golley, Frank Benjamin, J. T. McGinnis, R. G. Clements, G. I. Child, and M. J. Duever. "The Structure of Tropical Forests in Panama and Colombia." *BioScience* 19 (1969): 693–96.

Graham, J. B., I. Rubinoff, and M. K. Hecht. "Temperature Physiology of the Sea Snake *Pelamis platurus*: An Index of Its Colonization Potential in the Atlantic Ocean." *Proc. Nat. Acad. Sci. USA* 68 (1971): 1360–63.

Hanbury-Tenison, A. R., and P. J. K. Burton. "Should the Darien Gap Be Closed?" *Geographical Journal* 139 (1973): 43–52.

Härer, Andreas, Julián Torres-Dowdall, and Axel Meyer. "The Imperiled Fish Fauna in the Nicaragua Canal Zone." *Conservation Biology* 31 (2017): 86–95.

Haskin, Frederic J. *The Panama Canal*. Garden City, N.Y.: Doubleday, 1913.

Hayashi, Fumio, and Edward C. Prescott. "The 1990s in Japan: A Lost Decade." *Review of Economic Dynamics* 5 (2002): 206–35.

Heckadon-Moreno, Stanley. "Light and Shadows in the Management of the Panama Canal Watershed." In *The Rio Chagres: A Multidisciplinary Perspective of a Tropical River Basin*, edited by Russell S. Harmon, 28–44. Dordrecht: Springer, 2005.

Hildebrand, Samuel F. "The Panama Canal as a Passageway for Fishes, with List and Remarks on the Fishes and Invertebrates." *Zoologica* 24 (1939): 15–45.

Hubbs, Carl L. "Need for Thorough Inventory of Tropical American Marine Biotas before Completion of an Interoceanic Sea-Level Canal." In *1968 Symposium on Investigations and Resources of the Caribbean Sea and Adjacent Regions*, 467–70. Paris: UNESCO, 1971.

———, and T. E. B. Pope. "The Spread of the Sea Lamprey through the Great Lakes." *Transactions of the American Fisheries Society* 66 (1937): 172–76.

Huete-Pérez, Jorge A., Axel Meyer, and Pedro J. Alvarez. "Rethink the Nicaragua Canal." *Science* 347 (2015): 355.

———, Jose G. Tundisi, and Pedro J. Alvarez. "Will Nicaragua's Interoceanic Canal Result in an Environmental Catastrophe for Central America?" *Environmental Science and Technology* 47 (2013): 13217–19.

———, Manuel Ortega-Hegg, Gerald R. Urquhart, Alan P. Covich, Katherine Vammen, Bruce E. Rittmann, Julio C. Miranda, et al. "Critical Uncertainties and Gaps in the Environmental-and Social-Impact Assessment of the Proposed Interoceanic Canal through Nicaragua." *BioScience* 66 (2016): 632–45.

———, Pedro J. J. Alvarez, Jerald L. Schnoor, Bruce E. Rittmann, Anthony Clayton, Maria L. Acosta, Carlos E. M. Bicudo, et al. "Scientists Raise Alarms about Fast Tracking of Transoceanic Canal through Nicaragua." *Environmental Science and Technology* 49 (2015): 3989–96.

Humboldt, Alexander von. Alexander von Humboldt to Frederick M. Kelley, May 12, 1856. Reprinted in *Proceedings of the Royal Geographical Society of London* 1 (1857): 69–71.

———. *Cosmos: A Sketch of a Physical Description of the Universe.* Vol. 1. Translated by E. C. Otté. 1845; repr., London: Bell and Daldy, 1871.

———. *Political Essay on the Kingdom of New Spain.* Vols. 1, 2, and 4. Translated by John Black. London: Longman, 1814.

———. *Views of Nature.* Edited by Stephen T. Jackson and Laura Dassow Walls. Translated by Mark W. Person. Chicago: University of Chicago Press, 2014.

———, and Aime Bonpland. *Personal Narrative of Travels to the Equinoctial Regions of the New Continent, During the Years 1799–1804.* Vol. 6. Translated by Helen Maria Williams. 1821; repr., New York: AMS Press, 1966.

Inglis, David R., and Carl L. Sandler. "A Special Report on Plowshare—Prospects and Problems: The Nonmilitary Uses of Nuclear Explosives." *Bulletin of the Atomic Scientists* 23, no. 10 (1967): 46–53.

Johnson, Willis Fletcher. *Four Centuries of the Panama Canal.* New York: Henry Holt, 1906.

Jones, Meredith L., ed. "The Panamic Biota: Some Observations Prior to a Sea-Level Canal." *Bulletin of the Biological Society of Washington,* no. 2 (1972): 1–270.

———, and C. E. Dawson. "Salinity-Temperature Profiles in the Panama Canal Locks." *Marine Biology* 21 (1973): 86–90.

Kelley, F. M. "On the Connection between the Atlantic and Pacific Oceans, via the Atrato and Truando Rivers." *Proceedings of the Royal Geographical Society of London* 1 (1857): 63–69.

———. *The Union of the Oceans by Ship-Canal without Locks, via the Atrato Valley.* New York: Harper and Brothers, 1859.

Kelly, John S. "Moving Earth and Rock with a Nuclear Device." *Science* 138 (1962): 50–51.

Langer, Elinor. "ACDA: LBJ Supports Agency Plea for Bigger Budget, Longer Life; but Old Problems Still Remain." *Science* 147 (1965): 584–89.

———. "Project Plowshare: AEC Program for Peaceful Nuclear Explosives Slowed Down by Test Ban Treaty." *Science* 143 (1964): 1153–55.

Laycock, George. *The Diligent Destroyers.* New York: Doubleday, 1970.

Leopold, Aldo. *A Sand County Almanac and Sketches Here and There.* New York: Oxford University Press, 1949.

Leschine, Thomas M. "The Panamanian Sea-Level Canal: Problems and Prospects from a Policy Perspective." *Oceans 81* (1981): 615–19.

Linowitz, Sol M. *The Making of a Public Man: A Memoir.* Boston: Little, Brown, 1985.

Lloyd, John Augustus. "Account of Levellings Carried across the Isthmus of Panama, to Ascertain the Relative Height of the Pacific Ocean at Panama and of the Atlantic at the Mouth of the River Chagres." *Philosophical Transactions of the Royal Society of London* 120 (1830): 59–68.

Maass, Arthur. *Muddy Waters: The Army Engineers and the Nation's Rivers*. Cambridge, Mass.: Harvard University Press, 1951.

MacArthur Robert H., and Edward O. Wilson. *The Theory of Island Biogeography*. Princeton, N.J.: Princeton University Press, 1967.

Mahan, A. T. *The Influence of Sea Power upon History, 1660–1783*. 12th ed. Boston: Little, Brown, 1918.

———. "The Panama Canal and Sea Power in the Pacific." In *Armaments and Arbitration, or The Place of Force in the International Relations of States*, 155–80. New York: Harper and Brothers, 1912.

Mann, Roger, ed. *Exotic Species in Mariculture: Case Histories of the Japanese Oyster,* Crassostrea gigas *(Thunberg), with Implications for Other Fisheries*. Cambridge: Massachusetts Institute of Technology Press, 1978.

Martin, William E., U.S. Atomic Energy Commission, and Battelle Memorial Institute. *Symposium on Sea-level Canal Bioenvironmental Studies: Presented at the 19th Annual Meeting of the American Institute of Biological Sciences, September 4–5, 1968, at The Ohio State University*. Columbus, Ohio: Battelle Memorial Institute, 1969.

Mayr, Ernst. "Geographic Speciation in Tropical Echinoids." *Evolution* 8 (1954): 1–18.

McCosker, J. E., and C. E. Dawson. "Biotic Passage through the Panama Canal, with Particular Reference to Fishes." *Marine Biology* 30 (1975): 343–51.

McKeon, C. Seabird, Michele X. Weber, S. Elizabeth Alter, Nathaniel E. Seavy, Eric D. Crandall, Daniel J. Barshis, Ethan D. Fechter-Leggett, and Kirsten L. L. Oleson. "Melting Barriers to Faunal Exchange across Ocean Basins." *Global Change Biology* 22 (2015): 465–73.

Menzies, Robert J. "Transport of Marine Life between Oceans through the Panama Canal." *Nature* 220 (1968): 802–3.

Miller, A. Whitman, and Gregory M. Ruiz. "Arctic Shipping and Marine Invaders." *Nature Climate Change* 4 (2014): 413–16.

Moon, Katie, Vanessa M. Adams, Stephanie R. Januchowski-Hartley, Maksym Polyakov, Morena Mills, Duan Biggs, Andrew T. Knight, Edward T. Game, and Christopher M. Raymond. "A Multidisciplinary Conceptualization of Conservation Opportunity." *Conservation Biology* 28 (2014): 1484–96.

Mueller, Marti. "New Canal: What about Bioenvironmental Research?" *Science* 163 (1969): 165–67.

Muirhead, Jim R., Mark S. Minton, Whitman A. Miller, and Gregory M. Ruiz. "Projected Effects of the Panama Canal Expansion on Shipping Traffic and Biological Invasions." *Diversity and Distributions* 21 (2015): 75–87.

Murchison, R. I. *Address to the Royal Geographical Society of London*. London: W. Clowes and Sons, 1853.

Murphy, Arthur W. "The National Environmental Policy Act and the Licensing Process: Environmentalist Magna Carta or Agency Coup De Grâce?" *Columbia Law Review* 72 (1972): 963–1007.

New York Zoological Society. *1965 Annual Report*. New York: New York Zoological Society, 1966.

Newman, William A. "The National Academy of Science Committee on the Ecology of the Interoceanic Canal." In Jones, "The Panamic Biota," 247–59.

Nida, Stella Humphrey. *Panama and Its "Bridge of Water."* Chicago: Rand McNally, 1915.

O'Dea, Aaron, Harilaos A. Lessios, Anthony G. Coates, Ron I. Eytan, Sergio A. Restrepo-Moreno, Alberto L. Cione, Laurel S. Collins, et al. "Formation of the Isthmus of Panama." *Science Advances* 2, no. 8 (2016): e1600883. https://doi.org/10.1126/sciadv.1600883.

Porter, James W. "Ecology and Species Diversity of Coral Reefs on Opposite Sides of the Isthmus of Panama." In Jones, "The Panamic Biota," 89–116.

Reed, John C. "Ecological Investigation in the Arctic." *Science* 154 (1966): 372.

Reines, Frederick. "The Peaceful Nuclear Explosion." *Bulletin of the Atomic Scientists* 15, no. 3 (1959): 118–22.

Richards, P. W. "What the Tropics Can Contribute to Ecology." *Journal of Ecology* 51 (1963): 231–41.

Ripley, S. Dillon. "Perspectives in Tropical Biology." *BioScience* 17 (1967): 538–40.

Ripley, S. Dillon. *Trail of the Money Bird: 30,000 Miles of Adventure with a Naturalist*. New York: Harper and Brothers, 1942.

Ritterbush, Philip C. "Biology and the Smithsonian Institution." *BioScience* 17 (1967): 25–35.

Robinson, Nicholas A. "An Environmentalist Looks at the Panama Canal Treaties." *Sierra* (April 1978): 23–26.

———. "Extraterritorial Environmental Protection Obligations of Foreign Affairs Agencies: The Unfulfilled Mandate of NEPA." *New York University Journal of International Law and Politics* 7 (1974): 258–70.

———. "Introduction: Emerging International Environmental Law." *Stanford Journal of International Law* 17 (1981): 229–60.

Roosevelt, Theodore. "On American Motherhood." In Vol. 10 of *The World's Famous Orations*, edited by William Jennings Bryan and Francis W. Halsey, 253–62. New York: Funk and Wagnalls, 1906.

Ros, Macarena, Gail V. Ashton, Mariana B. Lacerda, James T. Carlton, Maite Vázquez-Luis, José M. Guerra-García, and Gregory M. Ruiz. "The Panama Canal and the Transoceanic Dispersal of Marine Invertebrates: Evaluation of the Introduced Amphipod *Paracaprella pusilla* Mayer, 1890 in the Pacific Ocean." *Marine Environmental Research* 99 (2014): 204–11.

Rosenblatt, Richard H. "The Zoogeographic Relationships of the Marine Shore Fishes of Tropical America." In *Proceedings of the International Conference on Tropical Oceanography*, 579–92. University of Miami Institute of Marine Sciences, 1967.

Rubinoff, Ira. "Central American Sea-Level Canal: Possible Biological Effects." *Science* 161 (1968): 857–61.

———. Letter to editor. *Science* 163 (1969): 762–63.
———. "Mixing Oceans and Species." *Natural History* 74, no. 7 (1965): 69–72.
———. "The Sea-Level Canal Controversy." *Biological Conservation* 3 (1970): 33–36.
———. "A Sea-Level Canal in Panama." In *The Tides of Change: Peace, Pollution, and Potential of the Oceans,* edited by Elisabeth Mann Borgese and David Krieger, 254–63. New York: Mason/Charter, 1975.
———, and Chaim Kropach. "Differential Reactions of Atlantic and Pacific Predators to Sea Snakes." *Nature* 228 (1970): 1288–90.
Rubinoff, Roberta W., and Ira Rubinoff. "Interoceanic Colonization of a Marine Goby through the Panama Canal." *Nature* 217 (1968): 476–78.
———. "Observations on the Migration of a Marine Goby through the Panama Canal." *Copeia,* no. 2 (1969): 395–97.
Ruiz, Gregory M., Mark E. Torchin, and Katharine Grant. "Using the Panama Canal to Test Predictions about Tropical Marine Invasions." *Smithsonian Contributions to the Marine Sciences* 38 (2009): 291–300.
Safriel, Uriel N. "The 'Lessepsian Invasion'—A Case Study Revisited." *Israel Journal of Ecology and Evolution* 59 (2014): 214–38.
Schlöder, Carmen, João Canning-Clode, Kristin Saltonstall, Ellen E. Strong, Gregory M. Ruiz, and Mark E. Torchin. "The Pacific Bivalve *Anomia peruviana* in the Atlantic: A Recent Invasion across the Panama Canal?" *Aquatic Invasions* 8 (2013): 443–48.
Seaborg, Glenn T. *Stemming the Tide: Arms Control in the Johnson Years.* Lexington, Mass.: D.C. Heath, 1987.
"The Second Panama Canal Project." *Japan Quarterly* 27 (1980): 303–7.
Sheffey, John P. "When Caribbean and Pacific Waters Mix." *Science* 162 (Dec. 20, 1968): 1329.
Smith, Scott A., Graham Bell, and Eldredge Bermingham. "Cross-Cordillera Exchange Mediated by the Panama Canal Increased the Species Richness of Local Freshwater Fish Assemblages." *Proceedings of the Royal Society of London B: Biological Sciences* 271 (2004): 1889–96.
Springer, Victor G. "Resolution on the Panama Canal." *Science* 182 (1973): 336.
Stein, Robert E. "United States Council on Environmental Quality Memorandum to U.S. Agencies on Applying the Environmental Impact Statement Requirement to Environmental Impacts Abroad." *International Legal Materials* 15 (1976): 1426–34.
Tate, Mercer D. "The Panama Canal and Political Partnership." *Journal of Politics* 25 (1963): 119–38.
———, and Edward H. Allen. "The Proposed New Treaties for the Panama Canal." *International Affairs* 45 (1969): 269–78.
Teller, Edward, Wilson K. Talley, Gary H. Higgins, and Gerald W. Johnson. *The Constructive Uses of Nuclear Explosives.* New York: McGraw-Hill, 1968.

Topp, Robert W. "Interoceanic Sea-Level Canal: Effects on the Fish Faunas." *Science* 165 (1969): 1324–27.

Torres de Araúz, Reina. *Demographic Characteristics of Human Groups Inhabiting the Eastern Region of the Republic of Panama*. Columbus, Ohio: Battelle Memorial Institute, 1968.

Travis, Martin B., and James T. Watkins. "Control of the Panama Canal: An Obsolete Shibboleth?" *Foreign Affairs* (1959): 407–18.

Vermeij, Geerat J. *Biogeography and Adaptation: Patterns of Marine Life*. Cambridge, Mass.: Harvard University Press, 1978.

Voss, Gilbert L. "Biological Results of the University of Miami Deep-Sea Expeditions." In Jones, "The Panamic Biota," 49–58.

———. "Panama Sea-Level Canal—II: Biological Catastrophe or Grand Experiment?" *Sea Frontiers* 24 (1978): 206–13.

Walsh, John. "National Academy of Sciences: Awkward Moments at the Meeting." *Science* 172 (1971): 539–42.

Wilson, Edward O., and E. O. Willis. "Applied Biogeography." In *Ecology and Evolution of Communities*, edited by Martin L. Cody and Jared M. Diamond, 522–34. Cambridge, Mass.: Harvard University Press, 1975.

Woodring, W. P. "The Panama Land Bridge as a Sea Barrier." *Proceedings of the American Philosophical Society* 110 (1966): 425–33.

Secondary Sources

Adas, Michael. *Dominance by Design: Technological Imperatives and America's Civilizing Mission*. Cambridge, Mass.: Harvard University Press, 2006.

Adler, Antony. *Neptune's Laboratory: Fantasy, Fear, and Science at Sea*. Cambridge, Mass.: Harvard University Press, 2019.

Andrews, Richard N. L. *Managing the Environment, Managing Ourselves: A History of American Environmental Policy*. 2nd ed. New Haven, Conn.: Yale University Press, 2006.

Anthony, Patrick. "Mining as the Working World of Alexander von Humboldt's Plant Geography and Vertical Cartography." *Isis* 109 (2018): 28–55.

Arntz, Dee. "Fletcher, Kathy (b. 1950)." HistoryLink.org Essay 9369. March 28, 2010.

Aron, William I., and Francis Dov Por. "A Tribute to Heinz Steinitz (1909–1971) and Gunnar Thorson (1906–1971)." *Israel Journal of Zoology* 21 (1972): 129–30.

Aronova, Elena, Karen S. Baker, and Naomi Oreskes. "Big Science and Big Data in Biology: From the International Geophysical Year through the International Biological Program to the Long Term Ecological Research (LTER) Network, 1957–Present." *Historical Studies in the Natural Sciences* 40 (2010): 183–224.

Balf, Todd. *The Darkest Jungle: The True Story of the Darién Expedition and America's Ill-Fated Race to Connect the Seas*. New York: Crown Publishers, 2003.

Balogh, Brian. *Chain Reaction: Expert Debate and Public Participation in American Commercial Nuclear Power, 1945–1975*. Cambridge: Cambridge University Press, 1991.

Billick, Ian, and Mary V. Price, eds. *The Ecology of Place: Contributions of Place-Based Research to Ecological Understanding*. Chicago: University of Chicago Press, 2012.

Blu Buhs, Joshua. *The Fire Ant Wars: Nature, Science, and Public Policy in Twentieth-Century America*. Chicago: University of Chicago Press, 2004.

Bocking, Stephen. *Ecologists and Environmental Politics: A History of Contemporary Ecology*. New Haven: Yale University Press, 1997.

———. "Situated but Mobile: Examining the Environmental History of Arctic Ecological Science." In *New Natures: Joining Environmental History with Science and Technology Studies*, edited by Dolly Jørgensen, Finn Arne Jørgensen, and Sara B. Pritchard, 164–78. Pittsburgh: University of Pittsburgh Press, 2013.

———, and Daniel Heidt, eds. *Cold Science: Environmental Knowledge in the North American Arctic during the Cold War*. New York: Routledge, 2019.

Bourne, Peter G. *Jimmy Carter: A Comprehensive Biography from Plains to Post-Presidency*. New York: Scribner, 1997.

Bowler, Peter J. *The Earth Encompassed: A History of the Environmental Sciences*. New York: W. W. Norton, 1992.

Brady, Scott. "An Historical Geography of the Earliest Colonial Routes across the American Isthmus." *Revista Geográfica* 126 (1999): 121–43.

Brannstrom, Christian. "Almost a Canal: Visions of Interoceanic Communication across Southern Nicaragua." *Ecumene* 2 (1995): 65–87.

Broderick, Mike. *Reconstructing Strangelove: Inside Stanley Kubrick's "Nightmare Comedy."* New York: Columbia University Press, 2017.

Brooks, Mark. "Economic Growth, Ecological Limits, and the Expansion of the Panama Canal." Ph.D. diss., McGill University, 2004.

Brower, David R. *For Earth's Sake: The Life and Times of David Brower*. Salt Lake City: Peregrine Smith Books, 1990.

Buys, Christian. "Isaiah's Prophecy: Project Plowshare in Colorado." *Colorado Heritage* 1 (1989): 28–39.

Caldwell, Lynton Keith. *The National Environmental Policy Act: An Agenda for the Future*. Bloomington: Indiana University Press, 1998.

Carlton, James T. "Blue Immigrants: The Marine Biology of Maritime History." *Log of Mystic Seaport Museum* 44 (1992): 31–36.

Carse, Ashley. *Beyond the Big Ditch: Politics, Ecology, and Infrastructure at the Panama Canal*. Cambridge, Mass.: MIT Press, 2014.

———. "An Infrastructural Event: Making Sense of Panama's Drought." *Water Alternatives* 10 (2017): 888–909.

———. " 'Like a Work of Nature': Revisiting the Panama Canal's Environmental History at Gatun Lake." *Environmental History* 21 (2016): 231–39.

———. "Nature as Infrastructure: Making and Managing the Panama Canal Watershed." *Social Studies of Science* 42 (2012): 539–63.

———, Christine Keiner, Pamela M. Henson, Marixa Lasso, Paul S. Sutter, Megan Raby, and Blake Scott. "Panama Canal Forum: From the Conquest of Nature to the Construction of New Ecologies." *Environmental History* 21 (2016): 206–87.

———, and David Kneas. "Unbuilt and Unfinished: The Temporalities of Infrastructure." *Environment and Society: Advances in Research* 10 (2019): 9–28.

Castro Herrera, Guillermo. "On Cattle and Ships: Culture, History and Sustainable Development in Panama." *Environment and History* 7 (2001): 201–17.

Caumartin, Corinne. Review of *Emperors in the Jungle*, by John Lindsay-Poland. *Journal of Latin American Studies* 36 (2004): 825–27.

Christen, Catherine A. "At Home in the Field: Smithsonian Tropical Science Field Stations in the U.S. Panama Canal Zone and the Republic of Panama." *The Americas* 58 (2002): 537–75.

Cittadino, Eugene. "Paul Sears and the Plowshare Advisory Committee." *Historical Studies in the Natural Sciences* 45 (2015): 397–446.

Clark, Ray, and Larry Canter, eds. *Environmental Policy and NEPA: Past, Present, and Future.* Boca Raton, Fla.: CRC Press, 1997.

Clayton, Lawrence A. "The Nicaragua Canal in the Nineteenth Century: Prelude to American Empire in the Caribbean." *Journal of Latin American Studies* 19 (1987): 323–52.

Clymer, Adam. *Drawing the Line at the Big Ditch: The Panama Canal Treaties and the Rise of the Right.* Lawrence: University Press of Kansas, 2008.

Coates, Anthony G., ed. *Central America: A Natural and Cultural History.* New Haven, Conn.: Yale University Press, 1997.

Coates, Peter. "Project Chariot: Alaskan Roots of Environmentalism." *Alaska History* 4 (1989): 1–31.

———. *The Trans-Alaska Pipeline Controversy: Technology, Conservation, and the Frontier.* Anchorage: University of Alaska Press, 1993.

Coen, Ross. *Breaking Ice for Arctic Oil: The Epic Voyage of the SS* Manhattan *through the Northwest Passage*. Fairbanks: University of Alaska Press, 2012.

Collins, George R. Introduction to *Unbuilt America: Forgotten Architecture in the United States from Thomas Jefferson to the Space Age*, edited by Alison Sky and Michelle Stone, 1–13. New York: McGraw-Hill, 1976.

Conniff, Michael L. *Black Labor on a White Canal: Panama, 1904–1981.* Pittsburgh: University of Pittsburgh Press, 1986.

Conniff, Michael L. *Panama and the United States: The End of the Alliance.* Athens: University of Georgia Press, 2012.

Conway, Erik. *High-Speed Dreams: NASA and the Technopolitics of Supersonic Transportation*. Baltimore: Johns Hopkins University, 2005.

Coulombe, Jordan. "Searching for Stability: Energy, Entropy, and the Abandoning of the Panatomic Canal." *Arcadia* (Rachel Carson Center for Environment and Society), no. 9 (Spring 2019). https://doi.org/10.5282/rcc/8506.

Covich, Alan P. "Frank Golley's Perspectives on Environmental Ethics and Literacy: How to Avoid Irreversible Impacts of Hydro-Power and Inter-Oceanic Canal

Development on Mesoamerican Tropical Ecosystems." *Human Ecology Review* 23, no. 2 (2017): 39–53.

———. "Projects That Never Happened: Ecological Insights from Darien, Panama." *Bulletin of the Ecological Society of America* 96 (2015): 54–63.

Cushman, Gregory T. "Humboldtian Science, Creole Meteorology, and the Discovery of Human-Caused Climate Change in South America." *Osiris* 26 (2011): 19–44.

d'Avignon, Robyn. "Shelf Projects: The Political Life of Exploration Geology in Senegal." *Engaging Science, Technology, and Society* 4 (2018): 111–30.

Davis, Jack E. *An Everglades Providence: Marjory Stoneman Douglas and the American Environmental Century*. Athens: University of Georgia Press, 2009.

Daynes, Byron W., and Glen Sussman. *White House Politics and the Environment: Franklin D. Roosevelt to George W. Bush*. College Station: Texas A&M University Press, 2010.

Delgado, James P., Tomás Mendizábal, Frederick H. Hanselmann, and Dominique Rissolo. *The Maritime Landscape of the Isthmus of Panamá*. Gainesville: University Press of Florida, 2016.

Doel, Ronald E. "Constituting the Postwar Earth Sciences: The Military's Influence on the Environmental Sciences in the USA after 1945." *Social Studies of Science* 33 (2003): 635–66.

———. "Scientists as Policymakers, Advisors, and Intelligence Agents: Linking Diplomatic History with the History of Science." In *The Historiography of the History of Contemporary Science, Technology, and Medicine*, edited by Thomas Söderqvist, 33–62. London: Harwood Academic Press, 1997.

———, and Kristine C. Harper. "Prometheus Unleashed: Science as a Diplomatic Weapon in the Lyndon B. Johnson Administration." *Osiris* 21 (2006): 66–85.

Donoghue, Michael E. *Borderland on the Isthmus: Race, Culture, and the Struggle for the Canal Zone*. Durham: Duke University Press, 2014.

———. "The Panama Canal and the United States." *Oxford Research Encyclopedia of American History*. Oxford: Oxford University Press, 2017. https://doi.org/10.1093/acrefore/9780199329175.013.260.

Dorsey, Kurkpatrick. "Dealing with the Dinosaur (and Its Swamp): Putting the Environment in Diplomatic History." *Diplomatic History* 29 (2005): 573–87.

———. *Whales and Nations: Environmental Diplomacy on the High Seas*. Seattle: University of Washington Press, 2013.

Dowie, Mark. *Losing Ground: American Environmentalism at the Close of the Twentieth Century*. Cambridge, Mass.: MIT Press, 1995.

Ealy, Lawrence O. *Yanqui Politics and the Isthmian Canal*. University Park: Pennsylvania State University Press, 1971.

Eastman, Adam R. "Hit List: President Carter's Review of Reclamation Water Projects and His Impact on Federal Water Policy." Ph.D. diss., University of Oklahoma, 2013.

Echenberg, Myron. *Humboldt's Mexico: In the Footsteps of the Illustrious German Scientific Traveller*. Montreal: McGill-Queen's University Press, 2017.

Egan, Michael. *Barry Commoner and the Science of Survival: The Remaking of American Environmentalism*. Cambridge, Mass.: MIT Press, 2007.

Eizenstat, Stuart E. *President Carter: The White House Years*. New York: St. Martin's Press, 2018.

Espeland, Wendy Nelson. *The Struggle for Water: Politics, Rationality, and Identity in the American Southwest*. Chicago: University of Chicago Press, 1998.

Ficek, Rosa E. "Imperial Routes, National Networks and Regional Projects in the Pan-American Highway, 1884-1977." *Journal of Transport History* 37 (2016): 129–54.

Findlay, Trevor. *Nuclear Dynamite: The Peaceful Nuclear Explosions Fiasco*. Sydney: Pergamon Press, 1990.

Finley, Carmel. *All the Fish in the Sea: Maximum Sustainable Yield and the Future of Fisheries Management*. Chicago: University of Chicago Press, 2011.

Fleming, James Rodger. *Fixing the Sky: The Checkered History of Weather and Climate Control*. New York: Columbia University Press, 2010.

Flippen, J. Brooks. *Conservative Conservationist: Russell E. Train and the Emergence of American Environmentalism*. Baton Rouge: Louisiana State University Press, 2006.

———. *Nixon and the Environment*. Albuquerque: University of New Mexico Press, 2000.

Flyvbjerg, Bent. "Survival of the Unfittest: Why the Worst Infrastructure Gets Built—and What We Can Do about It," *Oxford Review of Economic Policy* 25 (2009): 344–67.

Frenkel, Stephen. "Geography, Empire, and Environmental Determinism: The Case of Panama." *Geographical Review* 82 (1992): 143–53.

———. "A Hot Idea? Planning a Nuclear Canal in Panama." *Cultural Geographies* 5 (1998): 303–9.

Frisch, Scott A., and Sean Q. Kelly. *Jimmy Carter and the Water Wars: Presidential Influence and the Politics of Pork*. Amherst, N.Y.: Cambria Press, 2008.

Glad, Betty. *An Outsider in the White House: Jimmy Carter, His Advisors, and the Making of American Foreign Policy*. Ithaca, N.Y.: Cornell University Press, 2009.

Godbold, E. Stanly, Jr. *Jimmy and Rosalynn Carter: The Georgia Years, 1924–1974*. New York: Oxford University Press, 2010.

Goldin, Greg, and Sam Lubell. *Never Built Los Angeles*. New York: Metropolis Books, 2013.

———. *Never Built New York*. New York: Metropolis Books, 2016.

Golley, Frank Benjamin. *A History of the Ecosystem Concept in Ecology: More Than the Sum of the Parts*. New Haven, Conn.: Yale University Press, 1993.

Gonzalez, Carmen G. "Environmental Impact Assessment in Post-Colonial Societies: Reflections on the Proposed Expansion of the Panama Canal." *Tennessee Journal of Law and Policy* 4 (2008): 303–54.

Gottlieb, Robert. *Forcing the Spring: The Transformation of the American Environmental Movement*. Washington, D.C.: Island Press, 2005.

Graham, Loren R. *The Ghost of the Executed Engineer: Technology and the Fall of the Soviet Union*. Cambridge, Mass.: Harvard University Press, 1993.

Greenberg, David. *"What the Heck Are You Up To Mr. President?" Jimmy Carter, America's "Malaise," and the Speech That Should Have Changed the Country.* New York: Bloomsbury, 2009.

Greene, Benjamin P. *Eisenhower, Science Advice, and the Nuclear Test-Ban Debate, 1945–1963*. Stanford, Calif.: Stanford University Press, 2007.

Greene, Julie. *The Canal Builders: Making America's Empire at the Panama Canal.* New York: Penguin, 2009.

Haffer, Jürgen. *Ornithology, Evolution, and Philosophy: The Life and Science of Ernst Mayr, 1904–2005*. Berlin: Springer, 2007.

Hagen, Joel B. *An Entangled Bank: The Origins of Ecosystem Ecology.* New Brunswick, N.J.: Rutgers University Press, 1992.

———. "Problems in the Institutionalization of Tropical Biology: The Barro Colorado Island Biological Laboratory." *History and Philosophy of the Life Sciences* 12 (1990): 225–47.

Hamblin, Jacob Darwin. *Arming Mother Nature: The Birth of Catastrophic Environmentalism*. New York: Oxford University Press, 2013.

———. *Oceanographers and the Cold War: Disciples of Marine Science*. Seattle: University of Washington Press, 2005.

———. *Poison in the Well: Radioactive Waste in the Oceans at the Dawn of the Nuclear Age*. Piscataway, N.J.: Rutgers University Press, 2008.

Harper, Kristine C. *Make It Rain: State Control of the Atmosphere in Twentieth-Century America*. Chicago: University of Chicago Press, 2017.

Hays, Samuel P. *Conservation and the Gospel of Efficiency: The Progressive Conservation Movement, 1890–1920*. 1959; repr., Pittsburgh: University of Pittsburgh Press, 1999.

Hecht, Gabrielle, ed. *Entangled Geographies: Empire and Technopolitics in the Global Cold War.* Cambridge, Mass.: MIT Press, 2011.

Heffernan, Michael. "Bringing the Desert to Bloom: French Ambitions in the Sahara Desert during the Late Nineteenth Century—The Strange Case of 'La Mer Intérieure.' " In *Water, Engineering and Landscape: Water Control and Landscape Formation in the Modern Period*, edited by Denis Cosgrove and Geoff Petts, 94–114. London: Belhaven, 1990.

———. "Shifting Sands: The Trans-Saharan Railway." In *Engineering Earth: The Impacts of Megaengineering Projects*, edited by Stanley D. Brunn, 617–26. Dordrecht: Springer, 2011.

Helferich, Gerard. *Humboldt's Cosmos: Alexander von Humboldt and the Latin American Journey That Changed the Way We See the World.* New York: Penguin, 2004.

Henderson, Sandra L. "The Face of Empire: The Cultural Production of U.S. Imperialism in the Panama Canal Zone and California, 1904–1916." Ph.D. diss., University of Illinois at Urbana-Champaign, 2016.

Henson, Pamela M. "A Baseline Environmental Survey: The 1910–12 Smithsonian Biological Survey of the Panama Canal Zone." *Environmental History* 21 (2016): 222–30.

Hersey, Mark D., and Jeremy Vetter. "Shared Ground: Between Environmental History and the History of Science." *History of Science* 57 (2019): 403–40.

Hindle, Robert L. "Levees That Might Have Been." *Places Journal* (May 2015). https://placesjournal.org/article/levees-that-might-have-been/.

———. "Prototyping the Mississippi Delta: Patents, Alternative Futures, and the Design of Complex Environmental Systems." *Journal of Landscape Architecture* 12 (2017): 32–47.

Hogan, J. Michael. *The Panama Canal in American Politics: Domestic Advocacy and the Evolution of Policy*. Carbondale: Southern Illinois University Press, 1986.

———. "Public Opinion and American Foreign Policy: The Case of Illusory Support for the Panama Canal Treaties." *Quarterly Journal of Speech* 71 (1985): 302–17.

Howe, James. *Chiefs, Scribes, and Ethnographers: Kuna Culture from Inside and Out*. Austin: University of Texas Press, 2009.

———. *A People Who Would Not Kneel: Panama, the United States, and the San Blas Kuna*. Washington, D.C.: Smithsonian Institution Press, 1998.

Hunt, Jonathan. "Mexican Nuclear Diplomacy, the Latin American Nuclear-Weapon-Free Zone, and the NPT Grand Bargain, 1962–1968." In *Negotiating the Nuclear Non-Proliferation Treaty: The Making of a Nuclear Order*, edited by Andreas Wenger, Roland Popp, and Liviu Horovitz, 178–201. New York: Routledge, 2017.

Jaén Suárez, Omar. *Hombres y Ecología en Panamá*. Panamá: Editorial Universitaria, Smithsonian Tropical Research Institute, 1981.

Johnson, Kristin. "Natural History as Stamp Collecting: A Brief History." *Archives of Natural History* 34 (2007): 244–58.

Jones, Charles O. *The Trusteeship Presidency: Jimmy Carter and the United States Congress*. Baton Rouge: Louisiana State University Press, 1988.

Jorden, William J. *Panama Odyssey*. Austin: University of Texas Press, 1984.

Josephson, Paul R. *Industrialized Nature: Brute Force Technology and the Transformation of the Natural World*. Washington, D.C.: Island Press, 2002.

Kaufman, Burton I., and Scott Kaufman, *The Presidency of James Earl Carter*. Lawrence: University Press of Kansas, 2006.

Kaufman, Scott. *Plans Unraveled: The Foreign Policy of the Carter Administration*. DeKalb: Northern Illinois University Press, 2008.

———. *Project Plowshare: The Peaceful Use of Nuclear Explosives in Cold War America*. Ithaca, N.Y.: Cornell University Press, 2013.

Keiner, Christine. *The Oyster Question: Scientists, Watermen, and the Maryland Chesapeake Bay since 1880*. Athens: University of Georgia Press, 2009.

———. "A Two-Ocean Bouillabaisse: Science, Politics, and the Central American Sea-Level Canal Controversy." *Journal of the History of Biology* 50 (2017): 835–87.

Kingsland, Sharon E. *The Evolution of American Ecology, 1890–2000*. Baltimore: Johns Hopkins University Press, 2005.

Kinsey, Darin S. "'Seeding the Water as the Earth': The Epicenter and Peripheries of a Western *Aqua*cultural Revolution." *Environmental History* 11 (2006): 527–66.

Kirsch, David A. *The Electric Car and the Burden of History*. New Brunswick, N.J.: Rutgers University Press, 2000.

———. "Project Plowshare: The Cold War Search for a Peaceful Nuclear Explosive." In *Science, Values, and the American West*, edited by Stephen Tchudi, 191–222. Reno: Nevada Humanities Committee, 1997.

Kirsch, Scott. *Proving Grounds: Project Plowshare and the Unrealized Dream of Nuclear Earthmoving*. New Brunswick, N.J.: Rutgers University Press, 2005.

———, and Don Mitchell. "Earth-Moving as the 'Measure of Man': Edward Teller, Geographic Engineering, and the Matter of Progress." *Social Text*, no. 54 (1998): 100–34.

Knapp, Herbert, and Mary Knapp. *Red, White, and Blue Paradise: The American Canal Zone in Panama*. New York: Harcourt Brace Jovanovich, 1984.

Kohler, Robert E. *Landscapes and Labscapes: Exploring the Lab-Field Border in Biology*. Chicago: University of Chicago Press, 2002.

Kroll, Gary. *America's Ocean Wilderness: A Cultural History of Twentieth-Century Exploration*. Lawrence: University Press of Kansas, 2008.

Krygier, J. B. "Project Ketch: Project Plowshare in Pennsylvania." *Ecumene* 5 (1998): 311–22.

Kuhn, Thomas S. *The Structure of Scientific Revolutions*. Chicago: University of Chicago Press, 1962.

LaFeber, Walter. *The Panama Canal: The Crisis in Historical Perspective*. New York: Oxford University Press, 1978.

Laird, Frank N. *Solar Energy, Technology Policy, and Institutional Values*. New York: Cambridge University Press, 2001.

Lasso, Marixa. *Erased: The Untold Story of the Panama Canal*. Cambridge, Mass.: Harvard University Press, 2019.

Lawrence, Mark Atwood. "Exception to the Rule? The Johnson Administration and the Panama Canal." In *Looking Back at LBJ: White House Politics in a New Light*, edited by Mitchell B. Lerner, 20–47. Lawrence: University Press of Kansas, 2005.

Lehmann, Philipp Nicolas. "Infinite Power to Change the World: Hydroelectricity and Engineered Climate Change in the Atlantropa Project." *American Historical Review* 121 (2016): 70–100.

Lifset, Robert D. *Power on the Hudson: Storm King Mountain and the Emergence of Modern American Environmentalism*. Pittsburgh: University of Pittsburgh Press, 2014.

Lindsay-Poland, John. *Emperors in the Jungle: The Hidden History of the U.S. in Panama*. Durham, N.C.: Duke University Press, 2003.

———. "U.S. Military Bases in Latin America and the Caribbean." In *The Bases of Empire: The Global Struggle against U.S. Military Posts*, edited by Catherine Lutz, 71–96. New York: New York University Press, 2009.

Lindstrom, Matthew J., and Zachary A. Smith. *The National Environmental Policy Act: Judicial Misconstruction, Legislative Indifference, and Executive Neglect*. College Station: Texas A&M University Press, 2001.

Liroff, Richard A. "NEPA Litigation in the 1970s: A Deluge or a Dribble?" *Natural Resources Journal* 21 (1981): 315–30.

Li, Tania Murray. "Beyond 'the State' and Failed Schemes." *American Anthropologist* 107 (2005): 383–94.

Loo, Tina. "People in the Way: Modernity, Environment, and Society on the Arrow Lakes." *BC Studies*, nos. 142–143 (2004): 177–80.

Lutz, Catherine, ed. *The Bases of Empire: The Global Struggle against U.S. Military Posts*. New York: New York University Press, 2009.

Macekura, Stephen J. *Of Limits and Growth: The Rise of Global Sustainable Development in the Twentieth Century*. New York: Cambridge University Press, 2015.

Macfarlane, Daniel. "Negotiated High Modernism: Canada and the St. Lawrence Seaway and Power Project." In *Made Modern: Science and Technology in Canadian History*, edited by Edward Jones-Imhotep and Tina Adcock, 326–47. Vancouver: UBC Press, 2018.

Mack, Gerstle. *The Land Divided: A History of the Panama Canal and Other Isthmian Canal Projects*. New York: Knopf, 1944.

Major, John. *Prize Possession: The United States and the Panama Canal, 1903–1979*. Cambridge: Cambridge University Press, 1993.

Martin, Laura J. "Proving Grounds: Ecological Fieldwork in the Pacific and the Materialization of Ecosystems." *Environmental History* 23 (2018): 567–592.

Martini, Edwin A. *Agent Orange: History, Science, and the Politics of Uncertainty*. Amherst: University of Massachusetts Press, 2012.

Maurer, Noel, and Carlos Yu. *The Big Ditch: How America Took, Built, Ran, and Ultimately Gave Away the Panama Canal*. Princeton, N.J.: Princeton University Press, 2010.

Mayr, Ernst. *The Growth of Biological Thought: Diversity, Evolution, and Inheritance*. Cambridge, Mass.: Harvard University Press, 1982.

McCall, Sarah, and Matthew J. Taylor. "Nicaragua's 'Grand' Canal: Cuento Chino? Rhetoric and Field-Based Evidence on the Chinese Presence in Nicaragua." *Journal of Latin American Geography* 17 (2018): 191–208.

McCullough, David. *The Path between the Seas: The Creation of the Panama Canal, 1870–1914*. New York: Simon and Shuster, 1977.

McGuinness, Aims. *Path of Empire: Panama and the California Gold Rush*. Ithaca, N.Y.: Cornell University Press, 2008.

McKloskey, J. Michael. *In the Thick of It: My Life in the Sierra Club*. Washington, D.C.: Island Press, 2005.

McNamara, Robert S. *In Retrospect: The Tragedy and Lessons of Vietnam*. New York: Vintage Books, 1996.

McNeill, J. R., and Corinna R. Unger, eds. *Environmental Histories of the Cold War*. Washington, D.C.: Cambridge University Press, 2010.

McPherson, Alan. "Courts of World Opinion: Trying the Panama Flag Riots of 1964." *Diplomatic History* 28 (2004): 83–112.

———. "From 'Punks' to Geopoliticians: U.S. and Panamanian Teenagers and the 1964 Canal Zone Riots." *The Americas* 58 (2002): 395–418.

———. "Rioting for Dignity: Masculinity, National Identity, and Anti-U.S. Resistance in Panama." *Gender and History* 19 (2007): 219–41.

———. *Yankee, No! Anti-Americanism in U.S.–Latin American Relations*. Cambridge, Mass.: Harvard University Press, 2003.

Milam, Erika Lorraine. "The Equally Wonderful Field: Ernst Mayr and Organismic Biology." *Historical Studies in the Natural Sciences* 40 (2010): 279–317.

Millar, Susan W. S., and Don Mitchell. "Spectacular Failure, Contested Success: The Project Chariot Bioenvironmental Programme." *Ecumene* 5 (1998): 287–302.

Miller, Shawn W. "Minding the Gap: Pan-Americanism's Highway, American Environmentalism, and Remembering the Failure to Close the Darién Gap." *Environmental History* 19 (2014): 189–216.

Missal, Alexander. *Seaway to the Future: American Social Visions and the Construction of the Panama Canal*. Madison: University of Wisconsin Press, 2008.

Moffett, George D., III. *The Limits of Victory: The Ratification of the Panama Canal Treaties*. Ithaca, N.Y.: Cornell University Press, 1985.

Moore, Sarah J. *Empire on Display: San Francisco's Panama-Pacific International Exposition of 1915*. Norman: University of Oklahoma Press, 2013.

Morgan, Arthur E. *Dams and Other Disasters: A Century of the Army Corps of Engineers in Civil Works*. Boston: Porter Sargent, 1971.

Morris, Kenneth E. *Jimmy Carter: American Moralist*. Athens: University of Georgia Press, 1997.

Nelson, Derek Lee. "The Ravages of Teredo: The Rise and Fall of Shipworm in US History, 1860–1940." *Environmental History* 21 (2016): 100–124.

Newton, Velma. *The Silver Men: West Indian Labour Migration to Panama, 1850–1914*. Mona, Jamaica: University of the West Indies, 1984.

Niles, John M. *History of South America and Mexico; Comprising their Discovery, Geography, Politics, Commerce and Revolutions*. Vol. 2. Hartford, Conn.: H. Huntington, 1837.

Noll, Steven, and David Tegeder. *Ditch of Dreams: The Cross Florida Barge Canal and the Struggle for Florida's Future*. Gainesville: University Press of Florida, 2009.

Norse, Elliott A., and Larry B. Crowder. *Marine Conservation Biology: The Science of Maintaining the Sea's Biodiversity*. Washington, D.C.: Island Press, 2005.

Oatsvall, Neil. "Weather, Otters, and Bombs: Policy Making, Environmental Science, and U.S. Nuclear Weapons Testing, 1945–1958." In *Proving Grounds: Militarized Landscapes, Weapons Testing, and the Environmental Impact of U.S. Bases*, edited by Edwin A. Martini, 43–74. Seattle; London: University of Washington Press, 2015.

Oberdeck, Kathryn J. "Archives of the Unbuilt Environment: Documents and Discourses of Imagined Space in Twentieth-Century Kohler, Wisconsin." In *Archive Stories: Facts, Fictions, and the Writing of History*, edited by Antoinette Burton, 251–73. Durham, N.C.: Duke University Press, 2005.

O'Neill, Dan. *The Firecracker Boys: H-Bombs, Inupiat Eskimos, and the Roots of the Environmental Movement*. New York: St. Martin's Press, 1994.

———. "Project Chariot: How Alaska Escaped Nuclear Excavation." *Bulletin of the Atomic Scientists* 45, no. 10 (1989): 28–37.

Oreskes, Naomi. "Scaling Up Our Vision." *Isis* 105 (2014): 379–91.

Ovnick, Merry. "*Never Built Los Angeles*: A+D Architecture and Design Museum Los Angeles." *The Public Historian* 35 (2013): 73–76.

Parker, Matthew. *Panama Fever: The Epic Story of the Building of the Panama Canal.* New York: Anchor Books, 2009.

Parks, E. Taylor. *Colombia and the United States, 1765–1934*. Durham, N.C.: Duke University Press, 1935.

Pérez, Louis A., Jr. *The War of 1898: The United States and Cuba in History and Historiography*. Chapel Hill: University of North Carolina Press, 1998.

Peyton, Jonathan. *Unbuilt Environments: Tracing Postwar Development in Northwest British Columbia*. Vancouver: UBC Press, 2017.

Phillips, Fred M., G. Emlen Hall, and Mary E. Black. *Reining in the Rio Grande: People, Land, and Water*. Albuquerque: University of New Mexico Press, 2011.

Ponten, Josef. *Architektur, die nicht gebaut wurde* [Architecture that was not built]. Berlin: N.p., 1925.

Por, Francis Dov. "One Hundred Years of Suez Canal—A Century of Lessepsian Migration: Retrospect and Viewpoints." *Systematic Biology* 20 (1971): 138–59.

Pratt, Mary Louise. *Imperial Eyes: Travel Writing and Transculturation*. London: Routledge, 1992.

Primack, Joel R., and Frank Von Hippel. *Advice and Dissent: Scientists in the Political Arena*. New York: Basic Books, 1974.

Pritchard, Sara B. "Joining Environmental History with Science and Technology Studies: Promises, Challenges, and Contributions." In *New Natures: Joining Environmental History with Science and Technology Studies*, edited by Dolly Jørgensen, Finn Arne Jørgensen, and Sara B. Pritchard, 1–17. Pittsburgh: University of Pittsburgh Press, 2013.

Raby, Megan. *American Tropics: The Caribbean Roots of Biodiversity Science*. Chapel Hill: University of North Carolina Press, 2017.

———. "Ark and Archive: Making a Place for Long-Term Research on Barro Colorado Island, Panama." *Isis* 106 (2015): 798–824.

Ramey, Andrew. "The Calvert Cliffs Campaign, 1967–1971: Protecting the Public's Right to Knowledge." In *Nuclear Portraits: Communities, the Environment, and Public Policy*, edited by Laurel Sefton MacDowell, 121–49. Toronto: University of Toronto Press, 2017.

Rankin, William. "Zombie Projects, Negative Networks, and Multigenerational Science: The Temporality of the International Map of the World." *Social Studies of Science* 47 (2017): 353–75.

Rebok, Sandra. *Humboldt and Jefferson: A Transatlantic Friendship of the Enlightenment*. Charlottesville: University of Virginia Press, 2014.

Redfield, Peter. *Space in the Tropics: From Convicts to Rockets in French Guiana*. Berkeley: University of California Press, 2000.

Regis, Ed. *Monsters: The* Hindenburg *Disaster and the Birth of Pathological Technology*. New York: Basic Books, 2015.

Reisner, Marc. *Cadillac Desert: The American West and Its Disappearing Water*. 1986; repr., New York: Penguin, 1998.

Reuss, Martin. "The Art of Scientific Precision: River Research in the United States Army Corps of Engineers to 1945." *Technology and Culture* 40 (1999): 292–323.

———. "Seeing Like an Engineer: Water Projects and the Mediation of the Incommensurable." *Technology and Culture* 49 (2008): 531–46.

———, and Stephen H. Cutcliffe, eds. *The Illusory Boundary: Environment and Technology in History*. Charlottesville: University of Virginia Press, 2010.

Robertson, Thomas. *The Malthusian Moment: Global Population Growth and the Birth of American Environmentalism*. New Brunswick, N.J.: Rutgers University Press, 2012.

Rodgers, Ron. "From a Boon to a Threat: Print Media Coverage of Project Chariot, 1958–62." *Journalism History* 30 (2004): 11–19.

Roland, Alex, W. Jeffrey Bolster, and Alexander Keyssar. *The Way of the Ship: America's Maritime History Reenvisioned, 1600–2000*. Hoboken, N.J.: John Wiley and Sons, 2008.

Rome, Adam. *The Genius of Earth Day: How a 1970 Teach-In Unexpectedly Made the First Green Generation*. New York: Hill and Wang, 2013.

———. "What Really Matters in History? Environmental Perspectives on Modern America." *Environmental History* 7 (2002): 303–18.

Rothschild, Rachel. "Environmental Awareness in the Atomic Age: Radioecologists and Nuclear Technology." *Historical Studies in the Natural Sciences* 43 (2013): 492–530.

Rowe, Elana Wilson. "Promises, Promises: The Unbuilt Petroleum Environment in Murmansk." *Arctic Review on Law and Politics* 8 (2017): 3–16.

Rozwadowski, Helen M. "Engineering, Imagination, and Industry: Scripps Island and Dreams for Ocean Science in the 1960s." In *The Machine in Neptune's Garden: Historical Perspectives on Technology and the Marine Environment*, edited by Helen M. Rozwadowski and David K. Van Keuren, 315–53. Sagamore Beach, Mass.: Science History, 2004.

———. *Vast Expanses: A History of the Oceans*. London: Reaktion Books, 2018.

Rubinson, Paul. *Rethinking the American Antinuclear Movement*. New York: Routledge, 2018.

Rupke, Nicolaas. "A Geography of Enlightenment: The Critical Reception of Alexander von Humboldt's Mexico Work." In *Geography and Enlightenment*, edited by David N. Livingston and Charles W. J. Withers, 319–44. Chicago: University of Chicago Press, 1999.

Ryan, Paul B. *The Panama Canal Controversy: U.S. Diplomacy and Defense Interests*. Stanford, Calif.: Hoover Institution Press, 1977.

Rycroft, Robert W., and Joseph S. Szyliowicz. "Decision-Making in a Technological Environment: The Case of the Aswan High Dam." *World Politics* 33 (1980): 36–61.

Sachs, Aaron. *The Humboldt Current: Nineteenth-Century Exploration and the Roots of American Environmentalism*. New York: Penguin, 2006.

———. "The Ultimate 'Other': Post-Colonialism and Alexander von Humboldt's Ecological Relationship with Nature." *History and Theory* 42, no. 4. (2003): 111–35.

Sandars, Christopher. *America's Overseas Garrisons: The Leasehold Empire*. New York: Oxford University Press, 2000.

Sapp, Jan. *Coexistence: The Ecology and Evolution of Tropical Biodiversity*. Oxford: Oxford University Press, 2016.

———. *What Is Natural? Coral Reef Crisis*. Oxford: Oxford University Press, 1999.

Schwarz, Ingo. "Alexander von Humboldt's Visit to Washington and Philadelphia, His Friendship with Jefferson, and His Fascination with the United States." *Northeastern Naturalist* 8 (2001): 43–56.

Scott, James C. *Seeing Like a State: How Certain Schemes to Improve the Human Condition Have Failed*. New Haven, Conn.: Yale University Press, 1999.

Scoville, Caleb. "Hydraulic Society and a 'Stupid Little Fish': Toward a Historical Ontology of Endangerment." *Theory and Society* 48 (2019): 1–37.

Shallat, Todd. *Structures in the Stream: Water, Science, and the Rise of the US Army Corps of Engineers*. Austin: University of Texas Press, 1994.

Shor, Elizabeth N., Richard H. Rosenblatt, and John D. Isaacs. "Carl Leavitt Hubbs, 1894–1979." *National Academy of Sciences Biographical Memoirs* 56 (1987): 215–49.

Smith, Jason W. *To Master the Boundless Sea: The U.S. Navy, the Marine Environment, and the Cartography of Empire*. Chapel Hill: University of North Carolina Press, 2018.

Spear, Sheldon. *Daniel J. Flood: A Biography; The Congressional Career of an Economic Savior and Cold War Nationalist*. Bethlehem, Pa.: Lehigh University Press, 2008.

Spears, Ellen Griffith. *Rethinking the American Environmental Movement Post-1945*. New York: Routledge, 2019.

Stine, Jeffrey K. "Environmental Policy during the Carter Presidency." In *The Carter Presidency: Policy Choices in the Post–New Deal Era*, edited by Gary M. Fink and Hugh Davis Graham, 179–201. Lawrence: University Press of Kansas, 1998.

———. *Mixing the Waters: Environment, Politics, and the Building of the Tennessee-Tombigbee Waterway*. Akron, Ohio: University of Akron Press, 1993.

———, and Joel A. Tarr. "At the Intersection of Histories: Technology and the Environment." *Technology and Culture* 39 (1998): 601–40.

Strong, Robert A. "Jimmy Carter and the Panama Canal Treaties." *Presidential Studies Quarterly* 21 (1991): 269–86.

Suisman, David. "The American Environmental Movement's Lost Victory: The Fight against Sonic Booms." *The Public Historian* 37 (2015): 111–31.

Suman, Daniel O. "Socioenvironmental Impacts of Panama's Trans-Isthmian Oil Pipeline." *Environmental Impact Assessment Review* 7 (1987): 227–46.

Sutter, Paul S. "Nature's Agents or Agents of Empire? Entomological Workers and Environmental Change during the Construction of the Panama Canal." *Isis* 98 (2007): 724–54.

———. "Tropical Conquest and the Rise of the Environmental Management State." In *Colonial Crucible: Empire in the Making of the Modern American State*, edited by Alfred W. McCoy and Francisco A. Scarano, 317–26. Madison: University of Wisconsin Press, 2009.

———. "The World with Us: The State of American Environmental History." *Journal of American History* 100 (2013): 94–119.

Szylvian, Kristin M. "Transforming Lake Michigan into the 'World's Greatest Fishing Hole': The Environmental Politics of Michigan's Great Lakes Sport Fishing, 1965–1985." *Environmental History* (2004): 102–27.

Taylor, Serge. *Making Bureaucracies Think: The Environmental Impact Statement Strategy of Administrative Reform*. Stanford, Calif.: Stanford University Press, 1984.

Turner, Tom. *David Brower: The Making of the Environmental Movement*. Oakland: University of California Press, 2015.

Velásquez Runk, Julie. "Creating Wild Darien: Centuries of Darien's Imaginative Geography and Its Lasting Effects." *Journal of Latin American Geography* 14 (2015): 127–56.

Vetter, Jeremy, ed. *Knowing Global Environments: New Historical Perspectives on the Field Sciences*. Piscataway, N.J.: Rutgers University Press, 2010.

Walls, Laura Dassow. *The Passage to Cosmos*. Chicago: University of Chicago Press, 2011.

Weart, Spencer R. *The Rise of Nuclear Fear*. Cambridge, Mass.: Harvard University Press, 2012.

Weinstein-Bacall, Stuart. "The Darien Gap Case: Can Mere Words Interfere with the Sovereignty of a Foreign Nation?" *Lawyer of the Americas* 10 (1978): 589–608.

White, Richard. *The Organic Machine: The Remaking of the Columbia River*. New York: Hill and Wang, 1995.

Wilson, Edward O. *Nature Revealed: Selected Writings, 1949–2006*. Baltimore: Johns Hopkins University Press, 2006.

Wilt, Gloria, with Bart Hacker. "Gifts of a Fertile Mind." *Science and Technology Review* (July/Aug. 1998): 10–21.

Wolfe, Audra J. *Competing with the Soviets: Science, Technology, and the State in Cold War America*. Baltimore: Johns Hopkins University Press, 2013.

Wulf, Andrea. *The Invention of Nature: Alexander von Humboldt's New World*. New York: Knopf, 2015.

Zaretsky, Natasha. "Restraint or Retreat? The Debate over the Panama Canal Treaties and U.S. Nationalism after Vietnam." *Diplomatic History* 35 (2011): 535–62.

Zierler, David. *The Invention of Ecocide: Agent Orange, Vietnam, and the Scientists Who Changed the Way We Think about the Environment*. Athens: University of Georgia Press, 2011.

Zipp, Samuel. *Manhattan Projects: The Rise and Fall of Urban Renewal in Cold War New York*. New York: Oxford University Press, 2010.

INDEX

Jimmy Carter, the Politics of Family, and the Rise of the Religious Right,
by J. Brooks Flippen

Rumor, Repression, and Racial Politics: How the Harassment of Black Elected Officials Shaped Post–Civil Rights America,
by George Derek Musgrove

Doing Recent History: On Privacy, Copyright, Video Games, Institutional Review Boards, Activist Scholarship, and History That Talks Back,
edited by Claire Bond Potter and Renee C. Romano

The Dinner Party: Judy Chicago and the Power of Popular Feminism, 1970–2007,
by Jane F. Gerhard

Reconsidering Roots: Race, Politics, and Memory,
edited by Erica L. Ball and Kellie Carter Jackson

Liberation in Print: Feminist Periodicals and Social Movement Identity,
by Agatha Beins

Pushing Back: Women of Color–Led Grassroots Activism in New York City,
by Ariella Rotramel

Remaking Radicalism:
A Grassroots Documentary Reader of the United States, 1973–2001,
edited by Dan Berger and Emily K. Hobson

Deep Cut: Science, Power, and the Unbuilt Interoceanic Canal,
by Christine Keiner

CPSIA information can be obtained
at www.ICGtesting.com
Printed in the USA
LVHW090341251120
672602LV00008B/569